NOVICE AND GENERAL CLASS AMATEUR LICENSE Q & A MANUAL

NOVICE AND GENERAL CLASS AMATEUR LICENSE Q & A MANUAL

MARVIN TEPPER

W1YCV

HAYDEN BOOK COMPANY, INC.

Rochelle Park, New Jersey

Library of Congress Cataloging in Publication Data

Tepper, Marvin.
Novice and general class amateur license Q & A manual.

1. Radio--Examinations, questions, etc.
2. Radio operators--United States. I. Title.
TK6554.T47 621.3841'076 75-45168
ISBN 0-8104-5586-2

Printed in the United States of America

1 2 3 4 5 6 7 8 9 PRINTING

76 77 78 79 80 81 82 83 84 YEAR

INTRODUCTION

In its formative stages this book was updated many times to include all the latest FCC changes such as the new subject arrangement of questions (A. Rules and Regulations, B. Radio Phenomena, etc.) and the change from a list of specific study questions to the new syllabus-style study guide. Most important, of course, is the proposed restructuring of the amateur radio license classes and privileges.

Restructuring, as proposed by the FCC and illustrated in Fig. A, will divide amateur operators into what the FCC refers to as two "series" of amateur license classes, Series A and Series B, both culminating in the Extra class license which will offer full operating privileges of both series. Series A will permit operation on all amateur frequencies up to 29 MHz (HF), Series B on all amateur frequencies above 29 MHz (VHF). As shown in Fig. A, anyone desiring to enter the ranks of amateur radio operators will be faced with a choice of operating in what are presently the more popular frequencies—those below 29 MHz (Series A)—or the higher frequencies—those above 29 MHz (Series B), in which the relatively new and popular FM radiotelephone activities are taking place.

Each series is divided into operator categories, and each category offers specific frequency and power privileges. In Series A, in ascending order, the Novice class is restricted to operating code only on specified frequencies, using up to 250 watts of input power. A General class licensee is allowed to operate on most Series A frequencies and can operate code, AM, SSB, or FM radiotelephone using up to 500 watts of input power. Advanced class provides the same operating privileges as General class but with additional operating frequencies and up to 2000 watts of peak envelope power. In Series B, in ascending order, the Communicator class is restricted to operating phone only on specified frequencies and can use up to 250 watts of input power. The Technician is allowed to operate on most Series B frequencies and can operate code, AM, SSB, or FM radiotelephone using up to 500 watts of input power. The Experimenter license provides the same operating privileges as Technician class but with additional operating frequencies and up to 2000 watts of peak envelope power. An Extra class licensee is allowed all operator privileges in both series.

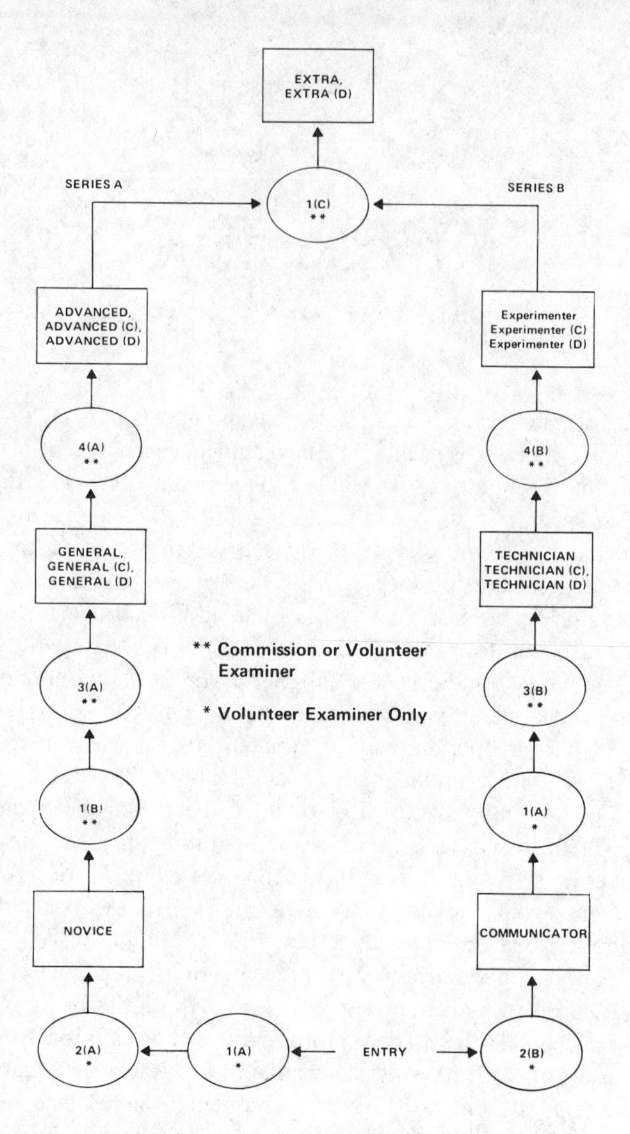

Fig. A. FCC-proposed restructuring of amateur licenses.

The encircled numbers in the illustration of the proposed structure indicate what tests (referred to as "Elements" by the FCC) must be taken to obtain a desired class of license. For example, an applicant for the Novice class must pass Element 1(A), a code test at 5 words per minute (wpm), and Element 2(A), a written multiple choice test on regulations and theory. For the Communicator license only a written test on regulations and theory is required. For the General, Technician, Advanced, and Experimenter

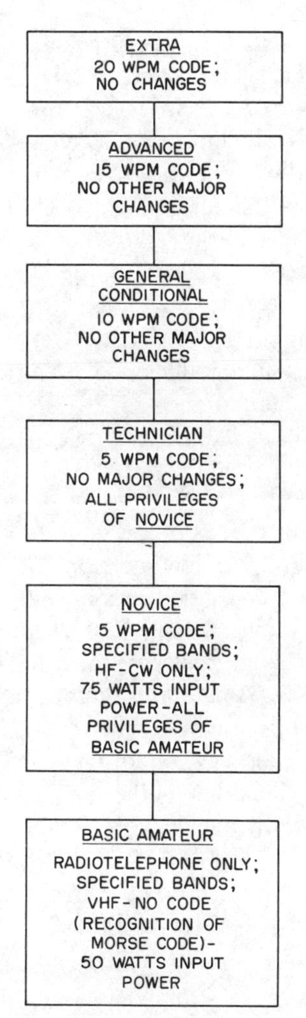

Fig. B. ARRL-proposed restructuring of amateur licenses.

licenses, all tests must be taken at an FCC office. The designations C and D indicate a conditionally issued license. Designator C is a temporary designation, for operators living at great distances from the nearest FCC office, or those who are out of the country or serving in the Armed Forces. Designator D indicates a conditional license for physically disabled persons. Restructuring also will allow an operator simultaneously to hold one license in Series A and one license in Series B, e.g., General and Technician licenses.

The American Radio Relay League (ARRL) in its reply to the FCC restructuring proposal has offered a series of counter-proposals. In brief, the ARRL favored retention of a single "ladder" or series of licenses, as shown in Fig. B. The ARRL concurred on a new no-code license class, called "Basic Amateur," in which operation would be restricted to VHF, although with less power than that provided for in the FCC-proposed Communicator class. The Novice license requirements are the same as those proposed by the FCC, but with the added privileges of a Basic Amateur so that the Novice could also operate phone on the VHF bands. The Technician license requirements also are the same, but with the added privileges of the Novice class, so a Technician licensee could also operate CW on the Novice portions of the HF bands. The other major changes proposed by the ARRL are in the reduced code speeds for General and Conditional classes, from 13 wpm to 10 wpm; and the increased code speed for the Advanced license, from 13 wpm to 15 wpm.

Regardless of the outcome, whether restructuring takes effect as proposed by the FCC or is modified by ARRL counter-proposals, it is a certainty that *some* form of restructuring will take effect. This book allows for any eventuality. The questions and answers in this book allow for the proposed FCC restructuring to take full effect. Possible modifications of the proposed restructuring are also covered in that any modifications will most likely be made in such areas as power requirements, operator class titles (Basic Amateur or Communicator), and operating privileges; thus the technical questions and answers, whether for Basic Amateur, Communicator, Novice, etc., will remain unchanged.

This book has been written expressly to help the new radio amateur or the amateur who wants to upgrade his license class. The author wishes all readers good luck in their exciting and rewarding endeavors in the amateur radio service.

Marvin Tepper

Falmouth, Massachusetts

CONTENTS

Chapter 1

GENERAL INFORMATION

History of Amateur Radio

In the earliest days of radio, little distinction existed between amateur and professional since radio was in its experimental stage and the body of knowledge about what was then called the "wireless" was both limited and available to everyone who desired it. The field soon became specialized, however. As radio was used by commercial and marine services, and the armed forces, a separate category of "experimenters" resulted and they quickly became known as amateur or "ham" operators.

By 1912, licensing of amateur radio operators and stations was under the Department of Commerce. The continental United States was divided into nine geographic areas and each amateur station was assigned a call sign starting with the numeral of the particular geographic area followed by the letters AA to AZ, and then through the alphabet to the letters WZ; for example, 1AW, 2PF, 9WZ, and so on. The letter X following the numeral was assigned to experimental stations, the letter Y to schools and colleges, and the letter Z to special authorized amateur stations; for example, 1XM, 3YA, and 9ZN.

All amateur radio operation ceased during World War I, and, with the help of the American Radio Relay League (ARRL), amateur radio was resumed after the war in October 1919. With the then new "audion" vacuum tube, and with the new short-wave frequencies [200 m (meters) and below, or 1500 kHz and above] assigned to amateurs to be investigated, amateur radio grew and its achievements became legend. Amateurs led the way to the use of higher radio frequencies, originally considered useless, and today amateurs must fight to preserve space in these "useless" frequencies.

In 1927, as a result of international treaties, different nations were assigned various call letters. The United States prefixes for commercial and amateur radio stations were W and K. Combining the prefix W or K with the geographic area number provides designators of W1 through W9 or K1 through K9. Call letters, from AA through ZZ, then AAA through ZZZ, were assigned. Thus, call signs start with W1AA, K1AA, and go through three initials to W9ZZZ, K9ZZZ. When all call letters through ZZZ are assigned, the letter A, then B, etc., follows the prefix; for example, WA1ABC, or KB6XYZ.

When World War II started, amateur radio was again suspended, and again amateur operators offered their specialized services to the nation's armed forces. After the war, the number of geographical areas in the United States was increased to ten (W0 being the new prefix) and the assignment of special meaning to the letters X, Y, and Z following the numeral was dropped. Amateur operators quickly resumed their pioneering. They pushed operation to still higher frequencies, bounced signals off the moon and off meteor-trails, operated through amateur satellites in space orbit, and most important, utilized new and improved methods and techniques, such as single sideband (SSB) and FM.

Although amateur radio operators have set an outstanding record, the majority are essentially hobbyists who appreciate and utilize the ability to communicate and make friends on a world-wide basis without leaving their homes. In late 1974, the FCC proposed a restructuring of the amateur radio service. The proposed changes are set forth in FCC Docket No. 20282, which is reprinted, courtesy of *Ham Radio Magazine,* Appendix III in this book.

Amateur Radio License Classes

Types of License Classes

To aid in acquiring the necessary skills and to reward the more proficient, the licensing of amateur radio operators has been divided into classes ranging from the basic Novice and proposed Communicator classes to the Extra class. These classes are listed below in their ascending order (the new proposed restructuring classes are noted by an asterisk):

- Novice
- Communicator*
- Technician
- (Conditional)†
- General
- Experimenter*
- Advanced
- Extra

Different tests are given by the Federal Communications Commission (FCC) for each class of license examination. The tests are most often referred to as examination elements. The elements are:

Element 1(A) – 5-wpm (words per minute) code test
Element 1(B) – 13-wpm code test
Element 1(C) – 20-wpm code test
Element 2(A) – Test on basic theory and FCC regulations for Novice class
Element 2(B) – Test on basic theory and FCC regulations for Communicator class

*New class of license under restructuring of FCC rules.
†The Conditional class license was originally offered as an alternative to the General class (see below). Under restructuring, there will be no Conditional class license per se, but the Technician, General, Experimenter, Advanced, and Extra class licenses may be obtained on a conditional basis.

Element 3(A) – Test on intermediate amateur practice and FCC regulations for General class
Element 3(B) – Test on intermediate amateur practice and FCC regulations for Technician class
Element 4(A) – Test on advanced radio theory and FCC regulations for Advanced class
Element 4(B) – (Present) Test on advanced radio theory and FCC regulations for Extra class
(Proposed) Test on advanced radio theory and FCC regulations for Experimenter class

The elements that must be passed to acquire each class of license are:

Novice – 1(A), 2(A)
Communicator – 2(B)
Technician – 1(A), 2(B), 3(B)
Conditional – 1(B), 2(A), 3(A)
General – 1(B), 2(A), 3(A)
Experimenter – 1(A), 2(B), 3(B), 4(B)
Advanced – 1(B), 2(A), 3(A), 4(A)
Extra – 1(C), 2(A), 2(B), 3(A), 3(B), 4(A), 4(B)

Under the proposed restructuring, the Conditional class per se will be eliminated, although Technician, General, Experimenter, Advanced, and Extra class licenses may be obtained conditionally.

The requirements and privileges of each class of amateur operation are listed in Table 1. The frequencies allocated for operation by each class of license are shown in Fig. 1. The privileges (code, kHz, MHz, etc.) are defined and discussed below.* The reader is advised to review this section of the book to learn the privileges accorded with each class of license.

The Novice class license was created to aid beginners in becoming proficient enough to obtain a General license. Novice requirements are minimal: a written test in elementary theory and FCC regulations and a 5-wpm code test. Passing these tests permits a Novice to operate only radiotelegraphy (CW) in specified amateur frequency bands. Maximum power input is presently 75 W (watts). The license is valid for 2 years and is not renewable. However, 1 year after the license has lapsed, the Novice test may be taken again, and a new Novice license will be issued for another 2-year period. Under the proposed restructuring, maximum power input would be 250 W. (This power rating is subject to change; be *certain* to check it when restructuring takes effect.) Also, the license will be effective for 5 years and renewable.

The Technician class license is expressly designed to encourage experimentation and development of the higher frequency amateur bands. Technician license requirements include the same test in theory and FCC regulations as given for the General class license, but the code speed is the same as that for a

*The use of the word cycles has been replaced by the name hertz to honor Heinrich Hertz, one of the first experimenters to envision radio "waves," or units of frequency. The terms that can be interchanged are: cycle = Hertz (Hz); kilocycle = kilohertz (kHz); and megacycle = megahertz (MHz).

Table 1(a). Amateur License Privileges.

Class	*Prior experience*	*Test requirements**	*Test elements*	*License term*	*Privileges*
Novice	None	By mail only	1(A), 2	2 years; Non-renewable; after a 1-year lapse, a new test may be taken	Code 3700–3750, 7100–7150, 21,100–21,200, and 28,100–28,200 kHz; maximum input power, 75 W; transmission type A1 (CW) only
Technician	None	By mail only	1(A), 3	5 years; renewable	All amateur privileges from 50.1–54 and 145–148 MHz; and all bands above 220 MHz
Conditional	None	By mail only	1(B), 3	5 years; renewable	All amateur privileges except those reserved for the Advanced and Extra class
General	None	In person by FCC examiner	1(B), 3	5 years; renewable	All amateur privileges except those reserved for the Advanced and Extra class
Advanced	None	In person by FCC examiner	1(B), 3, 4(A)	5 years; renewable	All amateur privileges except those reserved for the Extra class
Extra	1 year with a General or Advanced class license†	In person by FCC examiner	1(C), 3, 4(A), 4(B)	5 years; renewable	All amateur privileges

*By mail means that the test will be conducted by a volunteer examiner, as explained in Section 97.29(b).
†No test is given for holders of General or Advanced class licenses who can validate having held an amateur license prior to May 1917.

Table 1(b). Proposed Restructuring of Amateur License Privileges.

*Class**	*Test requirements and elements†*		*License term*	*Privileges*
	By mail only	*At FCC office*		
Novice	1(A), 2(A)		5 years; renewable	3700–3750, 7100–7150, 21,100–21,200, and 28,100–28,200 kHz; max. power input, 250 W; transmission type A1 (CW) only
Communicator	2(B)		5 years; renewable	All frequencies above 144 MHz; maximum power input, 250 W; transmission type F3 (FM) only
Technician (C) or (D)	1(A), 2(B)	3(B)	5 years; renewable	All frequencies above 50 MHz; maximum power input, 500 W; transmission type A1, A3, and F3
General (C) or (D)	2(A)	1(B), 3(A)	5 years; renewable	1800–2000, 3525–3775, 3890–4000, 7025–7150, 7225–7300, 14,025–14,200, 14,275–14,350, 21,025–21,250, and 21,350–21,450 kHz, and 28–29 MHz; maximum power input, 500 W; all transmission type privileges
Experimenter (C) or (D)	1(A), 2(B)	3(B), 4(B)	5 years; renewable	All frequencies above 29 MHz; maximum output, power, 2 kW PEP; all transmission type privileges
Advanced (C) or (D)	2(A)	1(B), 3(A), 4(A)	5 years; renewable	All frequencies below 29 MHz, except 3500–3525, 7000–7025, 14,000–14,025, and 21,000–21,025 kHz; maximum output power, 2 kW PEP; all transmission type privileges
Extra (D)	2(A) and (B)	1(C), 3(A) and (B), 4(A) and (B)	for life‡	All amateur radio operator privileges

*(C) and (D) designations are explained in Section 97.5(d) and (e) and Notice #17 of Proposed Rule Making in Appendix III.

†Elements 1(A), 2(A), and 2(B) are given by mail only, under the supervision of volunteer examiners (see Section 97.30 in Appendix III). All other test elements must be taken in person at an FCC office. If an applicant fails any element, the test may be retaken after 30 days (see Section 97.33 in Appendix III).

‡Operator privileges are granted for the life of the Extra class licensee; the station license must be renewed every 5 years.

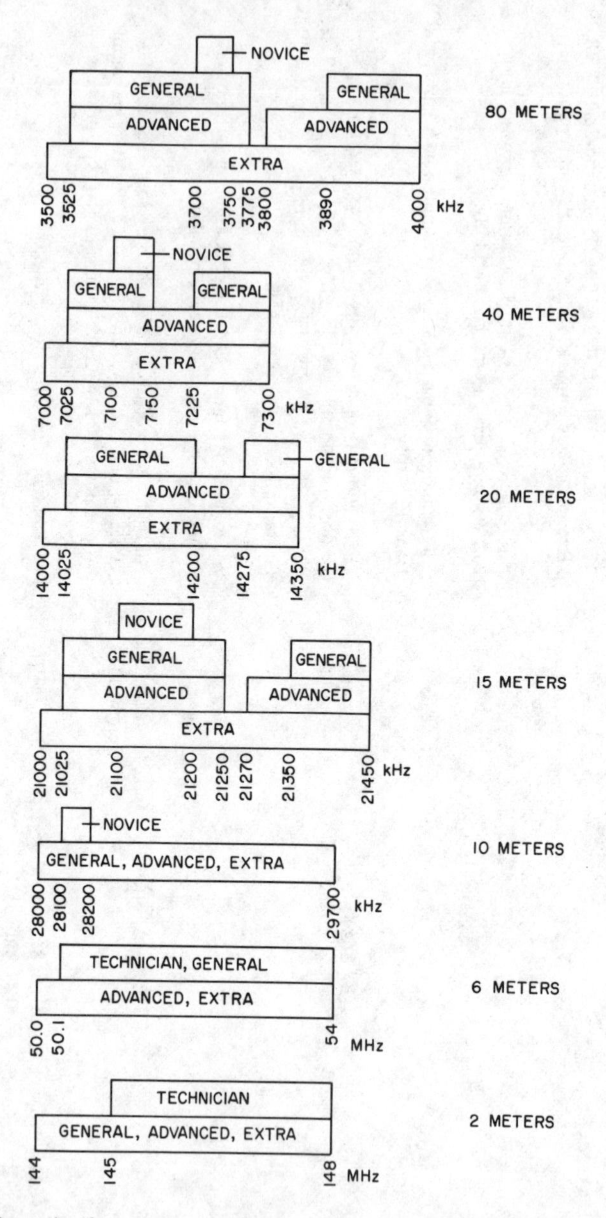

Fig. 1(a). Operating frequencies assigned to the different classes of amateur operators.

Novice, 5 wpm. The Technician license permits operation of all types: AM and FM, CW, MCW, Phone, TV, or Facsimile. Operation is permitted only on specified very-high-frequency bands.

The majority of radio amateurs hold the General class license. The theory test is of a level considered adequate for the operation and maintenance of a

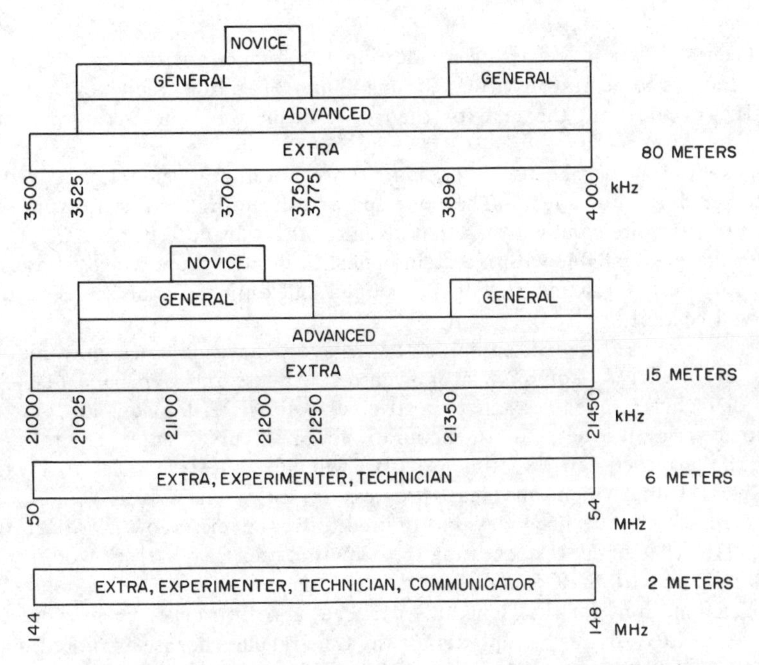

Fig. 1(b). Changes in frequency allocation by proposed restructuring.

radio amateur station. FCC Rules and Regulations must also be known. Code speed is a minimum of 13 wpm. A General class license permits operation of all types, as noted above for the Technician class license. There is presently no restriction on the bands available. Operation is allowed on portions of the 160-m band (as permitted in the given local area), the 2-m band, and the experimental higher frequencies above the 2-m band. Under the proposed restructuring, the maximum operating frequency would be 29 MHz.

The Conditional class license is the same as the General class license, but the examination is sent by mail and taken with a local volunteer examiner instead of being taken in an FCC office. To qualify for a Conditional license, the examinee must live beyond a 175-mile radius from the nearest FCC examining office, or have a physical disability, or be in the military service, or be a temporary resident outside of the United States or its territories for 1 year or more. The volunteer examiner must either hold an unexpired FCC amateur license (General, Advanced, or Extra) or an FCC commercial radiotelegraph license, or be the operator of a manually operated radiotelegraph station in the service of the United States. Under the proposed restructuring rules, there will no longer be a Conditional class license. Instead, all licenses other than Novice or Communicator may be obtained conditionally, although specific circumstances governing the issuance of these conditional licenses with volunteer examiners will be changed. (See the proposed amendment of Section 97.5(d) and Notice #17 of the Proposed Rule Making in Appendix III.)

The Advanced class license applicant must pass a test on advanced theory; no code test is required beyond that for a General class license. An Advanced class license holder may operate in additional frequencies within certain bands. Under the proposed restructuring, the maximum operating frequency would be 29 MHz. In addition, the test for theory would be the same as that presently required for Extra class.

The Extra class license is open to those holding a General or Advanced license for 1 or more years. The code speed test requirement is 20 wpm. The theory test is more complex than that given for the Advanced license, and added frequencies are available within certain bands. Under the proposed restructuring, the theory test requirements will be dropped and only the code speed requirement will be kept.

The proposed new Communicator class will have no telegraphy requirements or privileges. Communicator licensees will use the 2-m band (144-148 MHz) or higher frequencies and be restricted to F3 (FM) transmission. As with the Novice class, it is designed to encourage new amateurs to enter the ranks.

The proposed new Experimenter class will allow the Technician to advance to a class from which one may progress to Extra class. It is roughly the equivalent of the Advanced class but limited to frequencies above 29 MHz. Note in Fig. 1(b) that under restructuring the Experimenter class lisensee would share the 10-m band with the General and Advanced classes.

Examinations for the General, Advanced, and Extra class licenses must be taken at an FCC office. Exceptions for the General class license are noted under Conditional class license requirements. Distance and residence requirements are not allowable exceptions for Advanced and Extra class licenses. Physically disabled applicants for Advanced and Extra class licenses who cannot appear before an FCC examiner can, upon forwarding a physician's certificate, take the test at their own residences. The disability, however, must be long-term or permanent for the exception to apply. Under these circumstances, a volunteer may give the applicant the test. The volunteer must be 21 years of age or older, hold an amateur license equal to or of higher class than the test being given, or hold a commercial radiotelegraph operator's license issued by the FCC, or be the operator of a manually operated radiotelegraph station in the service of the United States. Under the proposed restructuring there will be new rules for volunteer examiners. In addition, the General, Technician, Experimenter, and Advanced licenses will be designated Type (C) or (D) when issued under special ("conditional") circumstances; refer to the proposed amendments of Sections 97.5(d) and (e), 97.27, and 97.28 and to Notice #17 of Proposed Rule Making in Appendix III.

Any element taken and passed in person at an FCC examining office need not be repeated for any other class of license. For example, if the examinee has passed Element 1(B) for General class license, he need not repeat that element for the Advanced class license.

However, any test taken by mail is not valid toward any other class of license. A holder of a Conditional class license must repeat all the tests in person at an FCC office to obtain the required elements for application for an Advanced class license. The holder of a Conditional class license trying for an Advanced class license must take all the tests in person. The tests can all be taken at one time; that is, the tests for Elements 1(B) and 3(A) for the General class can be followed with the test for Element 4(A) for the Advanced test during the same visit to the FCC office. Also, as noted in a subsequent section, *Filing Fees,* if all elements are taken during one visit, only one filing fee would be required.

Operator's License

For those who pass the FCC tests and receive a license, their license will be a single form but the license form is in two parts. Part 1, *Operator's Privileges,* gives the class of license, Novice, General, etc.; lists the call sign assigned; and gives the effective and expiration dates of the license.

Station License

Part 2 of the license form, *Station Location,* gives the address of the station if it differs from the operator's and lists the location of a remote control station, if any. Part 2 alone is often used by amateur clubs planning an amateur radio station that can be used by all its licensed members. Use of the club station is limited to operator's license privileges. The station operator, whether operating his own station or that of another person or group, must have his license with him when operating a station.

Another form of group-operated station is the "repeater" station used for high-frequency FM operation. The repeater station has a special station license using a call sign with the prefix WR.

FCC Examination Application Procedures

To help defray expenses, the FCC has instituted a filing fee which varies with the service offered. These fees are explained in detail later in this chapter, along with the locations of the various FCC engineering and field offices and their schedules for examinations. If the applicant can taken the examination at a regularly scheduled time in the nearest FCC field engineering office, the application form (see Fig. 2) can be filled out at the field office. If the applicant must take the examination at an examination point other than the field office, the application must be forwarded to the local field office. The FCC will then schedule an examination. If the applicant is taking the test by mail, Form 610 must first be received and filled out according to instructions as noted below.

Novice or Communicator License

To apply for a Novice or Communicator license, the applicant first must request a copy of Form 610 from the nearest FCC field engineering office. The applicant should read each question carefully and answer only those that pertain to the particular class application for which he is applying. At present, the Novice test is administered by a local volunteer examiner, who must have the following qualifications:

Be 21 years of age or older, and

Hold a valid FCC amateur license of either General, Advanced, or Extra class; or

Hold a valid FCC commercial radiotelegraph license; or

Currently operate a manually operated radiotelegraph station in the service of the United States.

FCC Form 610
July 1972

Form Approved
OMB No. 52-R0069

United States of America
Federal Communications Commission
Washington, D.C. 20554

APPLICATION FOR INDIVIDUAL AMATEUR RADIO STATION AND/OR OPERATOR LICENSE

Be sure to read instructions on reverse side.

FCC EXAMINER'S REPORT

FOR FCC USE ONLY.	ELE-MENT	1(A) SPEED	1(B) SPEED	1(C) SPEED	2	3	4(A)	4(B)
	PASSED							
	FAILED							

DATE ______ FCC EXAMINER ______ DIST. NO. ______

QUALIFIED FOR: ______

PLEASE TYPE OR PRINT

1. Name — Last, First, Middle Initial

2(a). Permanent Mailing Address — Number, Street; City, State, Zip Code

2(b). County:

3. Birthplace and Date — City, State, Month, Day, Year

4. Have you been convicted in a Federal, State, or local court of any crime for which the penalty imposed was a fine of $500 or more or an imprisonment of six months or more within ten years previous to the date of this application? (If "Yes," see Instruction B)
☐ YES ☐ NO

THIS APPLICATION IS FOR (Check appropriate boxes)

5. A. *New* Operator and/or Station License or *Change of Class* of Operator License (See Instruction C)
Check Class desired:
☐ Novice ☐ Technician ☐ Conditional
☐ General ☐ Advanced ☐ Amateur Extra

5. E. *Additional* Station License
For Present License give

Call Sign ______ Class Of Operator Privileges ______ Expiration Date ______

(NOTE: An additional station license always expires on the same date as the basic operator and station license) See Instruction D.

5. F *Modification* of Present License (See Instruction E)
Check modification desired;
☐ Change of permanent station location
☐ Other, Explain, using additional sheet if necessary.

5. G. *Renewal* of Present License (See Instruction F)
Have you met the operating time and code speed requirements of the Commission's Rules for renewal of operator license?
☐ YES ☐ NO

6. (a) I will appear for examination at ______
(See Instruction C(3)) City State

(b) Have you failed an amateur examination within the last 30 days?
☐ YES ☐ NO
If "Yes," state:
Class ______ Date of examination ______

ATTACH PRESENT LICENSE HERE—The present license or photocopy thereof must be attached unless it has been lost, mutilated, or destroyed, in which event the circumstances must be explained in this block.

Give the following information from present license

Station Call Sign ______ Class of Operator Privileges ______ Expiration Date ______

7. A. *Do you have any other amateur radio application on file* with the Commission that has not been acted upon?
☐ YES ☐ NO
B. If "Yes," state its purpose and date submitted.

8. If you request a special call sign, furnish information indicated in Instruction G
Special Call Sign Requested ______
If previously held, give Expiration Date ______

9. PRESENT or PROPOSED STATION LOCATION
If you do not have or do not propose to have a station, answer "None." Call letters are assigned only when a station location is given. (See Instruction H)
Number, Street
City, State
County

10. ANTENNA (all applicants must answer)
Will the antenna structure exceed an overall height of 200 feet above ground level or one foot above the established airport elevation for each 100 feet of distance, or fraction thereof, from the nearest boundary of such airport?
☐ YES ☐ NO
(If "Yes," complete and attach herewith FCC Form 714 obtainable at any Commission Office.)

11. If you propose to operate the transmitter by remote control, check whether control will be by WIRE ☐ or RADIO ☐ and attach your request for this authority. (See Instruction I)

12. CITIZENSHIP DECLARATION: See Instruction J.
☐ I am a citizen of the United States.
☐ I am an alien who has filed a Declaration of Intention to become a citizen of the United States.

WILLFUL FALSE STATEMENTS MADE ON THIS FORM ARE PUNISHABLE BY FINE AND IMPRISONMENT. U.S. CODE, TITLE 18, SECTION 1001.

DO NOT WRITE IN THIS BLOCK

N	Y	N	Y	N	Y	N	Y
CODE		EXAM		SCREEN		SIGN	

CERTIFICATION

I CERTIFY that all statements herein and attachments herewith are true, complete and correct to the best of my knowledge and belief and are made in good faith; that I am not the representative of any alien or of any foreign government; that I waive any claim to the use of any particular frequency or of the ether as against the regulatory power of the United States because of the previous use of the same whether by license or otherwise; that the station to be licensed, if any, will be inaccessible to unauthorized persons.

Applicant's Written Signature ______ Date Signed ______

Fig. 2. A portion of FCC Form 610.

Under the proposed restructuring, two volunteer examiners will be required; refer to the proposed amendment to Section 97.30 of Appendix III.

If the applicant cannot locate volunteer examiners in his area, he should write to the American Radio Relay League (ARRL), 225 Main Street, Newington, Conn. 06111. The League will try to locate a local volunteer examiner.

The volunteer examiner will first test the applicant's proficiency in code. Using equipment, such as a code oscillator, which will not transmit a radio signal, the examiner will transmit code at a speed of 5 wpm. Transmission will consist of five characters per word for a period of 5 minutes and will include numbers and punctuation (numbers and punctuation each count as two letters). To pass the test, it is necessary to copy 1 minute (25 successive characters) without error. Upon passing the receiving test, the applicant must then take the sending test. The required sending speed is no less than 5 wpm, and within a 5-minute period, at least 1 minute must be error free.

If the test is passed, the volunteer examiner will fill out the bottom of the reverse side of Form 610 and forward it to the FCC Licensing Unit, Gettysburg, Pa. 17325. This must be done within 10 days of the code examination. Upon receiving this form, the FCC will return to the volunteer examiner a set of papers containing the written examination for the Novice class license. The cover sheet of the test papers provides instructions for the examiner and the applicant. These instructions must be followed faithfully. The volunteer examiner will then administer the written test, which must be taken in his presence. The completed test must be returned to Gettysburg within 30 days of the date on which it was mailed from the FCC office. If the applicant has passed, he will receive a license in the mail within approximately 3 weeks. If the applicant has failed, he will be notified that the test was failed but not which questions were answered incorrectly.

A Novice or Communicator class license must be used with the privileges accorded to that class of license (see Table 1 and Fig. 1). The Novice license is presently valid for 2 years. If, by the end of those 2 years, the licensee has not advanced to a higher grade, the Novice license is no longer valid. It cannot be renewed, but a new exam can be taken 12 months after the expiration of the license. If the applicant fails the written test, he can file a new application 30 days (or later) after failing. He must retake the entire test. Under the proposed restructuring, the Novice and Communicator licenses will be valid for 5 years and will be renewable.

Technician License

The Technician license also is applied for by mail. All steps are identical to those of the Novice license application, including the code test requirements. The only exceptions are that Form 610 is filled out for a Technician class license and a filing fee is required when applying for the written examination. If the applicant for a Technician license holds a Novice license and passes the Technician test, his Novice license is canceled and he can use only the privileges of the Technician license.

A Technician class license must be used with the privileges accorded to that class of license (see Table 1 and Fig. 1). The license is good for 5 years and is continually renewable. If the written test is failed, a new application may be submitted 30 days (or later) after failing. The entire test must be retaken and a new fee paid.

General License

Most amateur radio operators hold the General class license. Applicants must take the test at an FCC office at scheduled times. Current locations and schedules are listed elsewhere in this book.

Before taking the test, the applicant must fill out Form 610 and pay the filing fee to the FCC examiner. The requirement for code proficiency is 13 wpm. The code test is given first. This is because the FCC has found through past experience that the majority of those who fail do so in the code test.

After passing the code test, the applicant takes the written examination. The examiner will grade the paper and let the applicant know whether he has passed or failed. If he has passed, he must wait about 3 weeks for his license to arrive in the mail. He *cannot* go on the air as a General class amateur until he receives the license.

A General class license must be used with the privileges accorded to that class of license (see Table 1 and Fig. 1). The license is good for 5 years and is continually renewable. If the test has been failed, a new application may be submitted 30 days (or later) after failing. The entire test must be retaken and a new fee paid.

Experimenter License

The proposed new Experimenter class license will allow the Technician to advance to a class from which one may progress to the Extra class license. The Experimenter will be permitted to operate any type of transmission on all frequencies above 29 MHz. The license will be valid for 5 years and renewable. Refer to Table 1(B) and Fig. 1(B) for a listing of the proposed privileges of the Experimenter license.

Conditional License

The Conditional license is identical to the General license, the only difference being that the applicant is unable to take the test in person at an FCC engineering office or field office. An applicant may take a Conditional license for the following reasons:

1. He lives more than 175 miles airline distance from the nearest location at which FCC examinations are held at intervals of not more than 6 months.
2. He can prove by a physician's certificate that he is unable to travel.
3. He is temporarily a resident outside the continental limits of the United States for a continuous period exceeding 1 year.
4. He is unable to appear at an FCC examination office because of military service as certified by the commanding officer.

As with the Novice license, the applicant for a Conditional license requests Form 610 from the nearest FCC field engineering office. The form is filled out for a Conditional class license, and a qualified volunteer examiner is located. A filing fee is required when applying for the written examination.

The code requirement for a Conditional class license is copying code at a speed of 13 wpm. Transmission consists of five characters per word for a period of 5 minutes and includes numbers, parentheses, etc. The applicant must be able to copy 1 minute (65 characters in succession) without an error to pass the test. Upon passing the receiving test, he must then take the code sending test. He is required to send at a speed of no less than 13 wpm and within a 5-minute period must send at least 1 minute (65 successive characters) without error.

A Conditional class license must be used with the privileges accorded to that class of license (see Table 1 and Fig. 1). The license is good for 5 years and is continually renewable. If the written test is failed, a new application may be submitted 30 days (or later) after failing. Again, the entire test must be retaken and a new fee paid.

If the Conditional class licensee should move to an area that is within the 175-mile distance limit, he need not reapply for a General class license but can renew the Conditional class license indefinitely.

When restructuring takes effect, Conditional licenses will be dropped, but General, Technician, Advanced, Experimenter, and Extra class will be awarded under conditional circumstances, as explained in Notice #17 of Proposed Rule Making and the proposed amendments of Section 97.5(d) and (e) of Appendix III.

Advanced and Extra Licenses

Advanced and Extra license tests must be taken at an FCC office, unless a physician's certificate states that the applicant cannot travel. A specially qualified volunteer examiner will then give the tests. The Advanced license requires a code speed of 13 wpm and a written test on intermediate radio theory and FCC regulations [Element 4(A)]. The Extra license requires a code speed of 20 wpm and a written test on advanced radio theory and FCC regulations [Element 4(B)].*

FCC Field Engineering Bureau Office Locations

There are presently 24 full-time FCC field engineering offices, which are located in major cities throughout the United States. They are listed on p. 14. Each city represents the main or central office for an individual FCC Field Engineering Bureau District. Figure 3 shows a map of the United States; the areas of each district are outlined and the district number indicated. In the listing of cities below, the FCC-assigned district numbers are also included.

---

*Detailed information regarding these two classes of license may be found in *Advanced and Extra Class Amateur License Q & A Manual,* also published by the Hayden Book Co., Rochelle Park, N.J.

FCC Field Engineering Offices

City	*District No.*	*City*	*District No.*
Anchorage, Alaska	23	Los Angeles, Calif.	11
Atlanta, Ga.	6	Miami, Fla.	7
Baltimore, Md.	4	New Orleans, La.	8
Boston, Mass.	1	New York, N.Y.	2
Buffalo, N.Y.	20	Norfolk, Va.	5
Chicago, Ill.	18	Philadelphia, Pa.	3
Dallas, Tex.	10	Portland, Ore.	13
Denver, Colo.	15	St. Paul, Minn.	16
Detroit, Mich.	19	San Francisco, Calif.	12
Honolulu, Hawaii	21	San Juan, P.R.	22
Houston, Tex.	9	Seattle, Wash.	14
Kansas City, Mo.	17	Washington, D.C.	24

There are also five branch offices at which amateur examinations are regularly held. Information regarding the examination schedule can be obtained from the main district office.

Branch Office	*Main Office*
Beaumont, Tex.	Houston, Tex.
Mobile, Ala.	New Orleans, La.
San Diego, Calif.	Los Angeles, Calif.
Savannah, Ga.	Atlanta, Ga.
Tampa, Fla.	Miami, Fla.

There are 31 cities in which examinations are held four times a year:

Albany, N.Y.
Birmingham, Ala.
Charleston, W. Va.
Cincinnati, Ohio
Cleveland, Ohio
Columbus, Ohio
Corpus Christi, Tex.
Davenport, Iowa
Des Moines, Iowa
Fort Wayne, Ind.
Fresno, Calif.
Grand Rapids, Mich.
Indianapolis, Ind.
Knoxville, Tenn.
Little Rock, Ark.
Louisville, Ky.
Memphis, Tenn.
Milwaukee, Wis.
Nashville, Tenn.
Oklahoma City, Okla.
Omaha, Nebr.
Phoenix, Ariz.
Pittsburgh, Pa.
St. Louis, Mo.
Salt Lake City, Utah
San Antonio, Tex.
Sioux Falls, S. Dak.
Syracuse, N.Y.
Tulsa, Okla.
Williamsport, Pa.
Winston-Salem, N.C.

There are 19 cities in which examinations are held twice a year:

Albuquerque, N. Mex.
Bangor, Maine
Boise, Idaho
El Paso, Tex.
Fairbanks, Alaska
Hartford, Conn.
Jackson, Miss.
Jacksonville, Fla.
Juneau, Alaska
Ketchikan, Alaska
Las Vegas, Nev.
Lubbock, Tex.
Portland, Maine
Reno, Nev.
Salem Va.
Spokane, Wash.
Tucson, Ariz.
Wichita, Kans.
Wilmington, N.C.

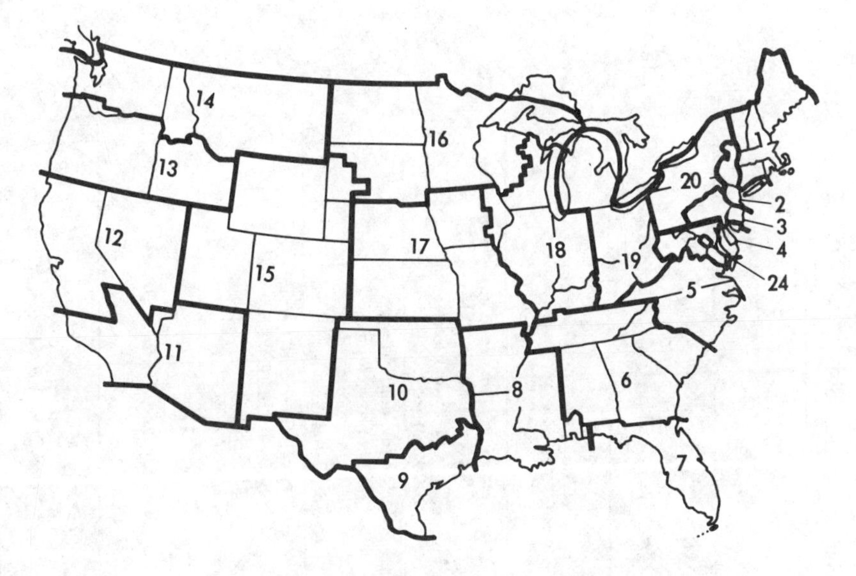

Fig. 3. FCC Field Engineering Bureau districts 1 through 20 and 24. Districts 21 (Hawaii), 22 (Puerto Rico), and 23 (Alaska) not shown.

In addition, FCC field engineers make yearly visits to the following cities:

Bakersfield, Calif.
Billings, Mont.
Great Falls, Mont.
Hilo, Hawaii
Helena, Mont.
Jamestown, N. Dak.
Klamath Falls, Ore.
Lihue, Hawaii
Marquette, Mich.
Rapid City, S. Dak.
Wailuku, Hawaii

FCC Examination Schedule

The current FCC examination schedule appears in Table 2. However, it is best to verify in advance the schedule and location of examinations being held in any FCC office since the FCC is trying to utilize offices of other branches of the government in cities that have no FCC office.

The applicant need not take the examination in the district in which he resides. If he happens to be in another district, he may take the examination at the scheduled times and places set by the main office of that district.

Filing Fees

With the exception of Novice class license, the FCC charges a $9 "filing" fee for each examination.* Every time the applicant takes an examination for a new license, as a result of having failed a previous examination, or for a higher grade license, the fee must be paid. This filing fee is for new licenses, renewals, or upgrading. The money is *not* returned if the examination is failed.

*As of this writing, there is a *proposed* increase in fees as follows: filing fee $10, modification fee $5 (Docket No. 19658).

Table 2. FCC District Office Locations and Examination Schedules.

Dist. no.	Office location	Examination schedule
1	BOSTON, MASSACHUSETTS 02109 1600 Customhouse India & State Streets *Phone: Area Code* 617 223-6608	P & ANC - Thursday and Friday 9:00 AM to 11:00 AM T & AC - Friday 9:00 AM to 11:00 AM
2	NEW YORK, NEW YORK 10014 748 Federal Building 641 Washington Street *Phone: Area Code* 212 620-5745	C & A - Tuesday through Thursday - 9:00 AM to 12:00 Noon
3	PHILADELPHIA, PENNSYLVANIA 19106 1005 U. S. Customhouse 2nd. & Chestnut Streets *Phone: Area Code* 215 597-4410	P & ANC - Monday, Tuesday, and Wednesday 10:00 AM to 12:00 Noon T & AC - Tuesday and Wednesday 8:00 AM to 9:00 AM
4	BALTIMORE, MARYLAND 21201 George M. Fallon Federal Building Room 819 31 Hopkins Plaza *Phone: Area Code* 301 962-2727	P - Monday and Friday 8:30 AM to Noon T & A - Monday and Friday 8:30 AM
5	NORFOLK, VIRGINIA 23502 Military Circle 870 North Military Highway *Phone: Area Code* 703 420-5100	P - Wednesday and Friday 9:00 AM to 12:00 Noon T & A - Thursday 9:00 AM ONLY
6	ATLANTA, GEORGIA 30303 1602 Gas Light Tower 235 Peachtree Street N. E. *Phone: Area Code* 404 526-6381	C & A - Tuesday and Friday 8:30 AM
6S	SAVANNAH, GEORGIA 31402 238 Federal Office Bldg. and Courthouse Bull and State Streets P. O. Box 8004 *Phone: Area Code* 912 232-4321 Ext. 320	P - Tuesdays - BY APPOINTMENT ONLY T & A - 2nd & 4th Monday of the Month BY APPOINTMENT ONLY

7	*MIAMI, FLORIDA 33130* *919 Federal Building* *51 S. W. First Avenue* *Phone: Area Code* 305 350-5541	P - Monday through Friday 8:00 a.m. to 4:30 p.m. T & A - Thursday 9:00 AM
7T	*TAMPA, FLORIDA 33606* *738 Federal Building* *500 Zack Street* *Phone: Area Code* 813 228-7711 Ext. 233	C & A - Tuesday through Friday 8:15 AM BY APPOINTMENT ONLY
8	*NEW ORLEANS, LOUISIANA 70130* *829 Federal Building South* *600 South Street* *Phone: Area Code* 504 527-2094	P & ANC - Tuesday and Wednesday 8:30 AM to Noon T & AC - Tuesday 8:30 AM
8M	*MOBILE, ALABAMA 36602* *439 U.S. Courthouse and Customhouse* *113 St. Joseph Street* *Phone: Area Code* 205 433-3581 Ext. 209	Please call Monday of the same week of exam for appointment C & A - Wednesday 8:00 AM BY APPOINTMENT ONLY
9	*HOUSTON, TEXAS 77002* *5636 Federal Building* *515 Rusk Avenue* *Phone: Area Code* 713 226-4306	P & ANC - Friday 8 AM to 12:00 Noon T & AC - Thursday 8 AM to 10:00 AM
9B	*BEAUMONT, TEXAS 77701* *323 Federal Building* *300 Willow Street* *Phone: Area Code* 713 838-0271 Ext. 317	P - Tuesday and Thursday BY APPOINTMENT ONLY T & A - Tuesday BY APPOINTMENT ONLY
10	*DALLAS, TEXAS 75202* *Room 13E7, 1100 Commerce St.* *Federal Building - U.S. Courthouse* *Phone: Area Code* 214 749-3243	P - Tuesday, Wednesday and Thursday 8:00 AM to 12:00 Noon T & A - Tuesday 8:00 AM to 12:00 Noon
11	*LOS ANGELES, CALIFORNIA 90012* *U. S. Courthouse, Room 1754* *312 N. Spring St.* *Phone: Area Code* 213 688-3276	P - Tuesday and Thursday 9:00 AM and 1:00 PM T & A - Wednesday 9:00 AM and 1:00 PM

(*continued*)

Table 2. FCC District Office Locations and Examination Schedules (*continued*).

Dist. no.	Office location	Examination schedule
11SD	SAN DIEGO, CALIFORNIA 92101 Fox Theatre Building 1245 Seventh Avenue *Phone: Area Code* 714 293-5460	P & ANC - Wednesday T & AC - First and Third Friday BY APPOINTMENT ONLY
11SP	SAN PEDRO, CALIFORNIA 90731 300 South Ferry Street Terminal Island *Phone: Area Code* 213 831-9281	Examinations are not normally conducted at San Pedro. Contact the FCC office at Los Angeles, California
12	SAN FRANCISCO, CALIFORNIA 94111 323A Customhouse 555 Battery Street *Phone: Area Code* 415 556-7700	P - Monday and Tuesday - 8:30 AM T - Tuesday - 8:30 AM A - Friday - Extra & Advanced (ANC) - 8:30 AM A - Friday - General & Advanced (AC) - 10:00 AM
13	PORTLAND, OREGON 97204 314 Multnomah Building 319 S.W. Pine Street *Phone: Area Code* 503 221-3097	P - Tuesday and Wednesday - 8:45 AM T - Tuesday - 8:45 AM A - Friday - 8:45 AM
14	SEATTLE, WASHINGTON 98104 8012 Federal Office Building 909 First Avenue *Phone: Area Code* 206 442-7653	P - Tuesday and Wednesday - 8:00 AM to 11:00 AM T & A - Friday - 8:45 AM
15	DENVER, COLORADO 80202 504 New Customhouse 19th St. between California & Stout Sts. Phone: Area Code 303 837-4054	P - Friday - 8:00 AM to 4:30 PM Tuesday through Thursday 9:00 AM or 12:30 PM by special appointment T & A - 1st & 2nd Thursdays of each month - 8:00 AM
16	ST. PAUL, MINNESOTA 55101 691 Federal Building 4th & Robert Streets *Phone: Area Code* 612 725-7819	C - Thursday - 8:45 AM A - Friday - 8:45 AM

17	KANSAS CITY, MISSOURI 64106 1703 Federal Building 601 East 12th St. *Phone: Area Code* 816 374-5526	C - Wednesday and Thursday Phone Third - 8:30 AM All other - 10:00 AM A - Thursday - 1:00 PM
18	CHICAGO, ILLINOIS 60604 1872 U. S. Courthouse 219 South Dearborn Street *Phone: Area Code 312 353-5386*	C - Thursday - 9:00 AM and 1:00 PM A - Friday - 9:00 AM
19	DETROIT, MICHIGAN 48226 1054 Federal Building Washington Blvd. & LaFayette Street *Phone: Area Code* 313 226-6077	C - Tuesday and Thursday - 9:00 AM A - Wednesday and Friday - 9:00 AM
20	BUFFALO, NEW YORK 14202 905 Federal Building 111 W. Huron St. at Delaware Ave. *Phone: Area Code* 716 842-3216	P - Thursday - 9:00 AM to 11:00 AM T & A - Friday - 9:00 AM Groups of 8 or more by appointment
21	HONOLULU, HAWAII 96808 502 Federal Building P.O. Box 1021 *Phone:* 546-5640	P - Monday, Thursday and Friday 8:00 AM, AND BY APPOINTMENT T & A - Tuesday and Wednesday 8:00 AM, AND BY APPOINTMENT
22	SAN JUAN, PUERTO RICO 00903 U. S. Post Office and Courthouse Room 322 - 323 P. O. Box 2987 *Phone: Area Code* 809 722-4562	P - Thursday and Friday - 8:30 AM T - Friday - 8:30 AM A - Friday - 9:00 AM
23	ANCHORAGE, ALASKA 99510 U. S. Post Office Building Room G63 4th & G Street P. O. Box 644 *Phone: Area Code 907* 272-1822	P - Monday through Friday - 8:00 AM to 3:00 PM T & A Monday through Friday BY APPOINTMENT ONLY
24	WASHINGTON, D. C. 20554 Room 216 1919 M Street, N.W. *Phone: Area Code* 202 632-7000	P - Tuesday and Friday 8:30 AM to 2:30 PM T & A - Friday 9:00 AM and 10:30 AM

Note: Schedule is subject to change. A = amateur exam; AC = amateur code exam; ANC = amateur code exam not required; C = commercial exam; P = radiotelephone exam; T = radiotelegraph exam.

Fees for services other than examinations include: modifications, such as a change of address, $4; a duplicate license, $6; and a request for a special call sign, $25.

The filing fee should be paid by check or money order made to the order of the Federal Communications Commission. No receipt will be given for checks or money orders received in the mail. If payment is made directly at an engineering office, a receipt will be furnished upon request.

Learning Code

Code Background

Code as a form of signaling is an extension of the ancient methods of using fire, smoke signals, drums, flags (semaphore), etc., to transmit messages. By assigning specific meanings to a combination of fires, sounds, flag colors, or flag positions, it is possible to say a lot with only a few symbols. In naval use, three colored pennants flying from a yardarm may indicate a complete message, such as "I am coming in to dock to refuel." Converting the code used for telegraph systems to a code useful for radio created the International Morse Code (Tables 3 and 4). The International Morse Code utilizes both a long and short duration of time, which becomes an audible sound at the receiver. A combination of sounds is assigned to each letter of the alphabet, to the numerals one through zero, to common punctuation marks, and to operating signals. Essentially, the code is simply another method of communicating, a language of its own, and, as in all languages, the more it is used the more proficient the user will become with the language, and the more he will enjoy using it.

Table 3. International Morse Code: Letters and Numbers.*

A	._	di-dah	S	...	di-di-dit
B	_...	dah-di-di-dit	T	_	dah
C	_._.	dah-di-dah-dit	U	.._	di-di-dah
D	_..	dah-di-dit	V	..._	di-di-di-dah
E	.	dit	W	.__	di-dah-dah
F	.._.	di-di-dah-dit	X	_.._	dah-di-di-dah
G	__.	dah-dah-dit	Y	_.__	dah-di-dah-dah
H	....	di-di-di-dit	Z	__..	dah-dah-di-dit
I	..	di-dit	1	.____	di-dah-dah-dah-dah
J	.___	di-dah-dah-dah	2	..___	di-di-dah-dah-dah
K	_._	dah-di-dah	3	...__	di-di-di-dah-dah
L	._..	di-dah-di-dit	4	...._	di-di-di-di-dah
M	__	dah-dah	5	.....	di-di-di-di-dit
N	_.	dah-dit	6	_....	dah-di-di-di-dit
O	___	dah-dah-dah	7	__...	dah-dah-di-di-dit
P	.__.	di-dah-dah-dit	8	___..	dah-dah-dah-di-dit
Q	__._	dah-dah-di-dah	9	____.	dah-dah-dah-dah-dit
R	._.	di-dah-dit	0	_____	dah-dah-dah-dah-dah

*From *Getting Started in Amateur Radio,* J. Berens and J. Berens. Rochelle Park, N.J.: Hayden Book Co., 1965.

Table 4. International Morse Code: Punctuation and Operating Symbols.

*From (DE)	– . . .	dah-di-dit dit
*End of message ($\overline{\text{AR}}$)†	. – . – .	di-dah-di-dah-dit
*End of work ($\overline{\text{SK}}$)	. . . – . –	di-di-di-dah-di-dah
*Wait ($\overline{\text{AS}}$)	. – . . .	di-dah-di-di-dit
*Invitation to transmit (K)	– . –	dah-di-dah
*Error	. . . .	di-di-di-di-di-di-di-dit
Understood ($\overline{\text{SN}}$)	. . . – .	di-di-di-dah-dit
*Received OK (R)	. – .	di-dah-di
*Period ($\overline{\text{AAA}}$)	. – . – . –	di-dah-di-dah-di-dah
Semicolon	– . – . – .	dah-di-dah-di-dah-dit
Colon	– – – . . .	dah-dah-dah-di-di-dit
Comma	– – . . – –	dah-dah-di-di-dah-dah
Quotes	. – . . – .	di-dah-di-di-dah-dit
*Question mark ($\overline{\text{IMI}}$)	. . – – . .	di-di-dah-dah-di-dit
Apostrophe	. – – – – .	di-dah-dah-dah-dah-dit
Hyphen	– . . . . –	dah-di-di-di-di-dah
Fraction bar	– . . – .	dah-di-di-dah-dit
Parentheses	– . – – . –	dah-di-dah-dah-di-dah
Underscore	. . – – . –	di-di-dah-dah-di-dah
*Double dash (break) ($\overline{\text{BT}}$)	– . . . –	dah-di-di-di-dah
Percent	. – – . – . – .	di-dah-dah-di-dah-di-dah-dit
Separation	. – . . –	di-dah-di-di-dah
*Attention ($\overline{\text{KA}}$)	– . – . –	dah-di-dah-di-dah

* The asterisk indicates the symbols required for the Novice exam.
†The bar above two letters indicates that they are run together.

Approach to Learning the Code

There is only one way to learn code: practice. Since the student cannot always have a code oscillator (practice set) with him, one of the simplest things he can do is to associate the so-called dots and dashes of the printed code with the phonetic sounds made as they are listened to through a receiver, *dit* and *dah*. The author used to study code by mentally reading the newspaper in code, for example, the word "and" became *dit-dah, dah-dit, dah-dit-dit,* and so on.

Practice does not mean round-the-clock; it is best to set aside a fixed amount of time each day, preferably at about the same time of day, and *stick to* that time. The student should not overdo practicing, even if he has time available. Practicing more than an hour is excessive; for most students, 30 to 45 minutes a day would suffice. However, the length of time and time of day chosen should remain constant.

An important principle of practicing is not to push. The student should not assume greater speed until he really feels comfortable at his current speed. When a student can copy and send confidently at the speed he is practicing and can consistently obtain 90 to 100% accuracy at that speed, he can consider himself "comfortable" at that speed. If he pushes, he will overreach; thinking he is proficient at some advanced speed and finding he cannot pass a test at that speed, he will be set back instead of making progress. This is sometimes called reaching a "plateau," described later in this chapter.

Why Code?

The simplest answer to why a code is necessary is simply to say that the law requires it. One reason the law requires code can be found in the name itself. *International* Morse Code. It is perhaps the only code in constant world-wide use. But, there are also technical reasons. When copying code versus speech, all things being equal, a weak code signal is much easier to copy than a weak speech signal. In addition, it takes less power at the code transmitter to put out a signal whose strength is equal to or greater than that of the speech signal. Also, there are human failings. A word transmitted in code has only the original meaning; the same word spoken can be misinterpreted; for example, was the message "*he* will visit you" or was it "*she* will visit you?" To sum up, code is in universal use, requires less power to send over greater distances, is easier to copy under difficult reception conditions, and contains no ambiguities in the message.

Code Key and Practice Oscillator

Figure 4 illustrates a typical hand-operated code key, in particular the adjustable portions of the key. The spring tension adjustment is a personal one. As a rule, the tension becomes lighter with increased experience and the ability to send at a faster speed. The space adjusting screw is the one most often adjusted. With time and experience, an operator will find a suitable gap spacing. The beginner should try a gap spacing of approximately 1/32 of an inch.

When using a key, it is best to mount the key on a heavy board, or to make provisions to clamp or screw the key (or mounting board) in place in order to prevent the key from moving as it is operated. The arm should rest on a table or other flat surface and the key should be held as shown in Fig. 5. When using

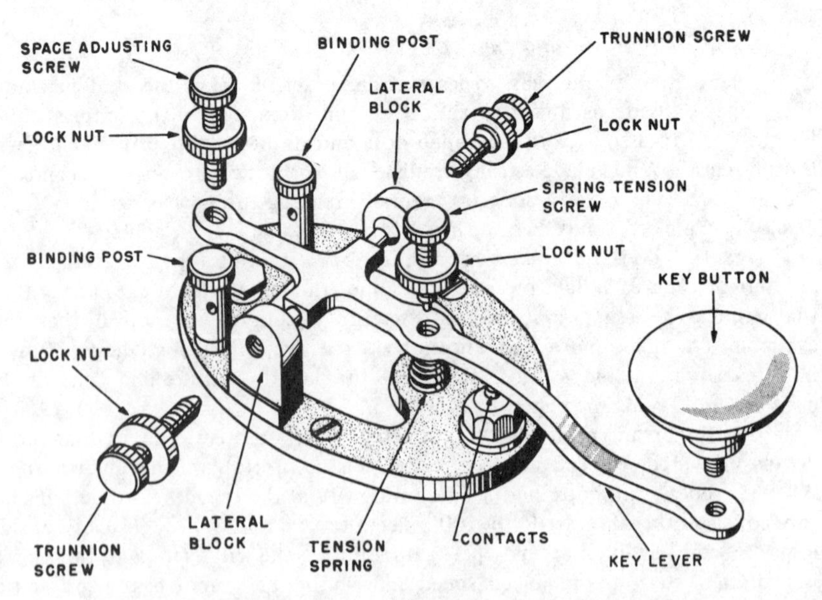

Fig. 4. A hand-operated code key. Courtesy, Wm. M. Nye Co., Inc.

Fig. 5. Proper position for operating the code key.

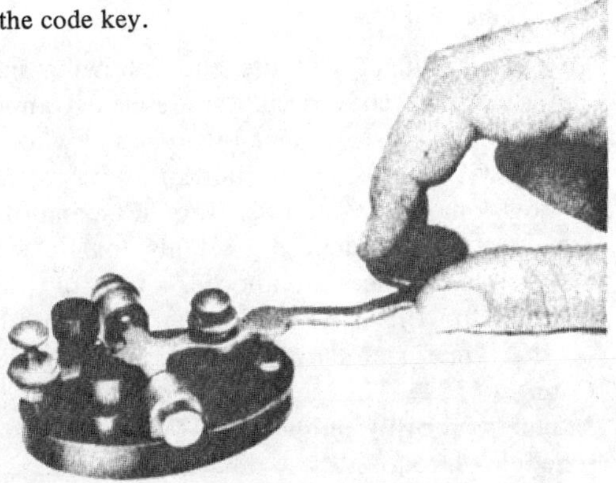

the key, the entire forearm should be used, not just the wrist. Being comfortable helps prevent fatigue.

Although the simplest listening device is a key-operated buzzer, it is perhaps best to try to duplicate the sound of code as it will be heard from a radio receiver. To do this requires an oscillator. Two simple and inexpensive code oscillators are shown in Fig. 6.

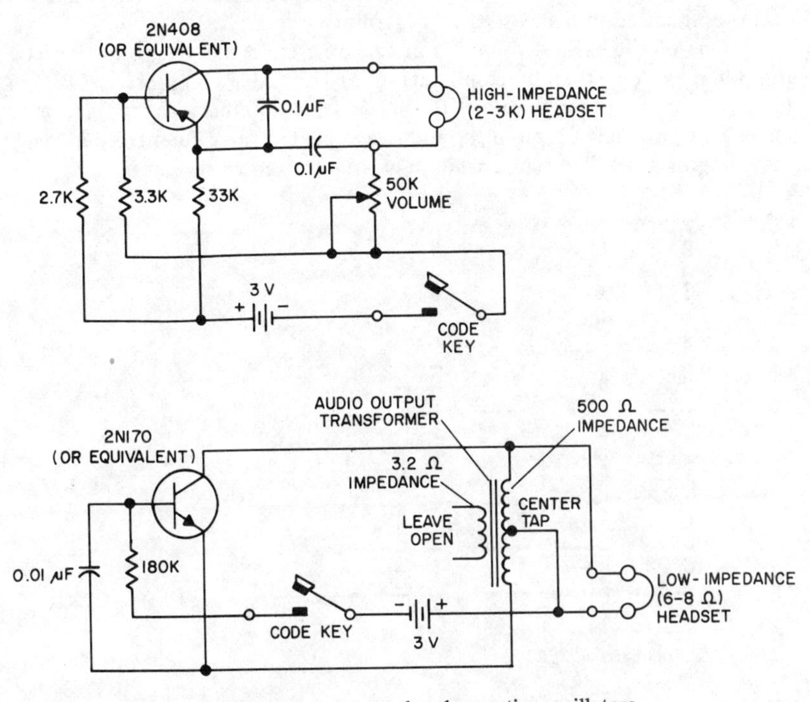

Fig. 6. Transistor-operated code practice oscillators.

Learning the Code

The student can use the following information to teach himself code, employing the code oscillator described above. Perhaps the best way to learn code is to listen to it sent by someone who is proficient. An amateur operator who is willing to spend some time transmitting code will give a beginner an idea of how long a *dit* and a *dah* should be and of just how code sounds at the speed for which the student has set his goal–5 wpm or 13 wpm. If it is difficult to obtain the services of another operator, the student should listen to other amateurs on the air. In particular, he should listen to station W1AW, the station of the American Radio Relay League (ARRL), 225 Main Street, Newington, Conn. 06111. The schedule for practice code transmission is listed in the League's monthly publication, *QST*. ARRL will provide a copy of the practice schedule on request.

If at all possible, the student should try to obtain the use of a code-sending machine, since the FFC uses this in the code test. Although expensive, code-sending machines may be available at rental rates in large cities. Another way to learn code is with phonograph records or tapes. One such record set is the *Sound-n-Sight Code Course* by Robins and Harris, (Hayden Book Company). This set combines flash cards and recordings to provide reinforced sound and sight learning.

There are various approaches to learning the code; the best known of these are discussed below. There are advocates of each method and no one method has been proved more effective than others. The student should select whichever one seems best suited to his own ease of learning.

One of the earliest methods of teaching code was to group the letters by *dits,* then by *dahs,* then by combinations of increasing complexity. This is shown in Fig. 7. In another method, the letters and numbers are in random order, instead of in similar groupings, and are repeated in different orders each time they are practiced. Two such random groups are shown in Fig. 8.

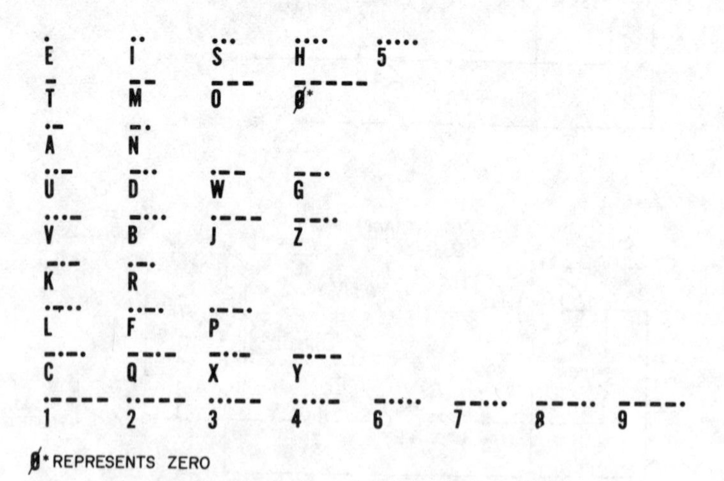

Fig. 7. "Increasing-complexity" scheme for teaching code.

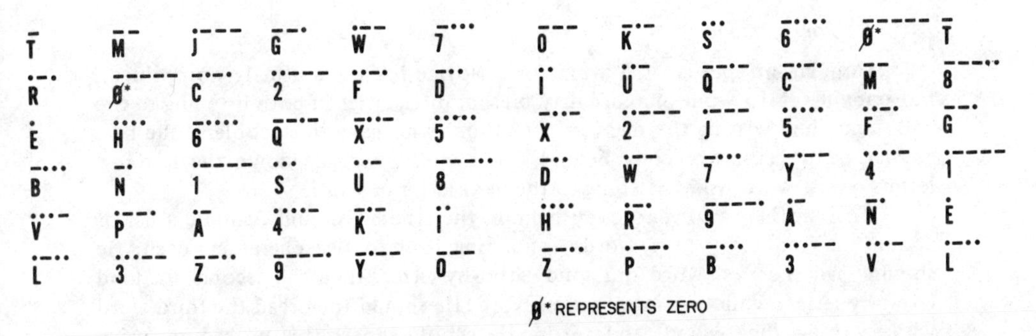

Fig. 8. "Random-order" schemes for teaching code.

Still another method is simply to memorize the code in alphabetical order, just as it is illustrated in Table 3. Another method is to learn the code by first memorizing all the characters starting with *dit,* then all those starting with *dah,* as shown in Fig. 9. (Note the ascending order of complexity of the characters.)

The Plateau

If the beginner talks to amateurs who have become proficient at code speeds of 13 wpm or more, he will hear references to a *plateau.* This concerns reaching a particular speed and then having extreme difficulty in increasing that speed. The plateau is often reached at about 10 wpm, and it is a common complaint of those studying for a General or Conditional license. Usually, the plateau exists only in the student's mind. What typically happens is that the student has reached "saturation," whether due to monotony, cramming, distraction, or more likely a change in study habits. Often this leads to a feeling that he can go no faster and "this is it," referred to as a plateau. This in turn is followed by a holiday from practice, which in turn reinforces the so-called plateau. However, if he follows one of the approaches to learning code described above, carefully and consistently, he will probably not reach a plateau in his learning.

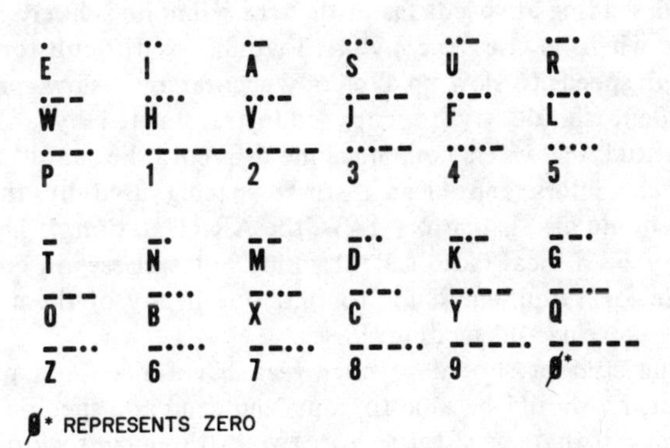

Fig. 9. "Dit/dah-separation" scheme for teaching code.

Copying Speed

Copying at speeds of 5 wpm for a Novice license is usually no problem, but copying at 13 wpm or more may present difficulties in both listening to the code and then writing the message. For those who have this problem, the best answer is to learn to "copy behind." This means writing down the last few letters or last word while listening to the next letter or word.

To help in learning to copy behind, the student should compile a list of about ten frequently used words containing four to five characters each. He should read the first word out loud letter-by-letter, then the second out loud letter-by-letter, while writing the first word. He should then read the third word while writing the second, and so on. He should repeat this procedure, using different lists, until the copying becomes reasonably easy. Next, the student should repeat the procedure, but instead of writing the word, he should send it by key on the practice oscillator.

After practice in transmitting behind, the next step would be to copy behind. At first try copying only one letter behind, no more. Then in time try two letters behind, then three, etc. As his confidence increases, he can copy behind an amount determined only by the time he spends practicing.

Taking the Code Examination

As noted earlier, applicants for a radio license are given the code test first since the majority of applicants who fail do so because of the code test. This puts the examinee under stress. It is well known that most people do not do their best when working under stress. The trick is to remove that stress, to approach the code test with the utmost confidence. The following suggestions are the best-known methods for developing confidence at copying and sending code: The code used in the test is sent by machine. It is excellent copy; that is, it is loud and clear with no background noise. Each character is precise.

The student should practice at home with standard type earphones if he can, since that is the type he will use to copy code during the test.

If the student is going to take the test for 13 wpm, he should practice until he can copy and send at approximately 15 to 18 wpm; *no faster.* If he practices at copying and sending at speeds faster than 18 wpm, he is likely to stumble and make mistakes when he tries to copy at 13 wpm. It is difficult for someone who can copy at high speeds to slow up and copy accurately at slow speeds.

The student should try to copy code transmitted by a machine or by records. When studying with a code machine or record, he should try to send the same precise characters and clear distinct spacing used by the machine or recording. He should also listen to W1AW, the ARRL station. If he lives in a large city, there may be a local radio school which holds classes in code. He should check with the local amateur clubs to find out if any of them have regularly scheduled code copying and sending classes.

Again, the student should practice *regularly* for a fixed period of time. Most important, he should be able to copy and send at a specific speed with 90 to 100% accuracy at that speed for at least two or three *consecutive* times.

Practicing copying at home should be done on standard 8½ X 11 ruled pads of paper. The same copying tool, pen or pencil, should be used all the time. If the student uses a pen, he should take it with him to the test since the FCC furnishes pencils only for the test.

The student should stick to the style of writing that suits him best, script (cursive) writing or printing. When transmitting, the key should be adjusted as closely as possible to the spacing and tension to which it was adjusted at home.

At this writing, the FCC is trying a new testing procedure which replaces code copying with code recognition: A short message is transmitted; after transmission, the applicant is given a sheet of paper on which the exact message transmitted and several similar transmissions are printed. The applicant selects the transmission he "recognizes" as the correct transmission.

On the day of the test, before leaving the house, the student should copy and transmit at 15 to 18 wpm. If he manages this with 100% accuracy at home for a full 5 minutes, he will know he has it made for the required 1-minute error-free portion of the test at the FCC office. That is confidence.

Learning the Technical Background

The specific purpose of this book is to prepare the student for the FCC examination for Novice and Technician, Conditional, or General class amateur radio tests. Although the FCC (as of this writing) provides a *Study Guide,* which lists typical questions asked on an FCC examination, there is no guarantee that the questions will be exactly the same as those given in the *Study Guide,* but they will be similar. The FCC is considering the use of a syllabus-type study guide in place of the question type; an FCC-type syllabus for the Novice class is shown in Fig. 10.

This book lists the type of questions covered by both the original study guide and the proposed syllabus-type study guide. Each numbered question (e.g., A.2.1) is followed by a brief but complete *Answer.* This is the type of answer the student will find in the FCC multiple-choice examination. To help the student understand the answers, additional background information has been presented where appropriate.

The background material for each question, listed as *Discussion,* is limited to the question being discussed. As a result, the student may find himself reading a portion of a discussion he does not understand. He is advised to learn all about radio communications electronics before taking the test. Since this is not always practical, it is recommended that the student at least learn what is required to pass the FFC examination. Reference books on the subject of radio communications circuits include: *Basic Radio,* M. Tepper, Hayden Book Co.; *The Radio Amateur's Handbook,* American Radio Relay League; *Understanding Amateur Radio,* American Radio Relay League; *Radio Handbook,* Editors and Engineers.

The student should study the questions, answers, and discussions. Where he finds he does not fully comprehend the answer or discussion, he should turn to a reference book and study that portion relating to what he does not understand.

Fig. 10. FCC-type Syllabus Study Guide for Element 2: Novice class amateur radio operator's license.

A. Rules and Regulations

A.2.1 *Basis and purpose:* Voluntary noncommercial communications service; Advancement of the radio art; Creation of a reservoir of trained radio operators and electronic experts.
A.2.2 *Definitions:* Amateur radio service; Amateur radio operator; Amateur radio station; Control operator; Station license; Primary station.
A.2.3 *Novice class operator privileges:* Frequencies; Emissions; Transmitter power.
A.2.4 *Limitations:* License period; Antenna structures.
A.2.5 *Responsibilities:* Station licensee; Control operator; Third party.
A.2.6 *Station operation:* Station identification; Station logs; Station license availability; Operator license availability; One-way communications; Points of communication; Frequency measurement.
A.2.7 *Administrative sanctions:* Notice of violation; Restricted operation.
A.2.8 *Prohibited practices:* Broadcasting; Interference; Unidentified communications; Third-party traffic.
A.2.9 *Licenses:* Operator; Station; Renewal.

B. Radio Phenomena

B.2.1 *Definitions:* Sky wave; Direct (free-space) wave; Ground wave; Surface wave; Skip distance; Wavelength; Ionosphere; Sunspot cycle.
B.2.2 *Wave characteristics:* Polarization, Spreading; Speed versus medium.
B.2.3 *Wave propagation:* Frequency versus distance; Wavelength versus frequency; Speed of radio waves; Affects of ionization upon wave propagation; Seasonal variations; Daylight versus night hours; Ionospheric storms; Sky wave versus ground wave.

C. Operating Procedures

C.2.1 *Basic principles:* Frequency selection.
C.2.2 *Telegraphy procedures:* Q-signals; Standard abbreviations; RST reporting system.
C.2.3 *Public service operating:* Responsibility; Message traffic; Network operation.

D. Emission Characteristics

D.2.1 *Definitions:* Continuous waves; Carrier frequency; Spurious emissions; Key clicks.
D.2.2 *Classification of emissions:* A0; A1
D.2.3 *Factors concerning A1 emissions:* Standards of good quality; Methods of keying; Frequency stability; Key clicks.

E. Electrical Principles

E.2.1 *Definitions:* Electromotive force; Direct current; Alternating current; Resistance; Capacitance; Inductance; Voltage drop; Electrical power; Rectification; Harmonics; Kilohertz; Megahertz.
E.2.2 *Fundamental units:* Volt; Ampere; Ohm; Watt, Henry; Farad.
E.2.3 *Direct current theory:* Ohms law; Resistance in series; parallel, and series-parallel; Power.

Fig. 10. FCC-type Syllabus Study Guide for Element 2: Novice class amateur radio operator's license (*continued*).

F. Practical Circuits

F.2.1 *Basic circuits:* Battery with internal resistance and external load; Basic amplifier circuit; Basic oscillator circuit.
F.2.2 *Rectifier circuits:* Half-wave; Full-wave.

G. Circuit Components

G.2.1 *Components:* Capacitors; Resistors; Inductors; Transformers; Crystals.
G.2.2 *Solid-state components:* Diodes; Transistors.
G.2.3 *Vacuum tubes:* Classification by elements.
G.2.4 *Meters:* Voltage meter; Current meter.

H. Antennas and Transmission Lines

H.2.1 *Definitions:* Electrical length for half-wave antenna; Field strength; Characteristic impedance; Standing waves; Transmatch.
H.2.2 *Half-wave dipole antenna:* Basic characteristics; Length versus frequency; Harmonic operation.
H.2.3 *Transmission lines:* Single conductor; Parallel conductor; Coaxial; Matching; Standing waves.

I. Radio Communication Practices

I.2.1 *Definitions:* Ground rod; Lightning protection.
I.2.2 *Radio interference:* Television interference; Radio interference.
I.2.3 *Use of test equipment:* Voltmeter; Ammeter; Ohmmeter.
I.2.4 *Transmitters:* Determination of input power; Determination of transmitter frequency.
I.2.5 *Safety:* Electrical shock avoidance. Lightning protection.

After he acquires his license, the student will find that while operating a station he will encounter technical questions that require learning new information. He will also acquire much technical background information while ragchewing with other radio amateurs.

Review: The first step is to study the questions, answers, and discussions carefully. If the student doesn't understand a part of an answer or discussion, he should consult a reference book and study only the part that pertains to the question or discussion. After he has his license, he should learn as much as possible about all of the radio communications electronics necessary for amateur radio.

Taking the Written Examination

The FCC written test is multiple choice. For each question asked, five possible answers (marked A through E) are given, only one of which is correct. The examination questions are printed in groups of 50 questions each and numbered accordingly: 1–50, 51–100, 101–150, and so on. An examinee may receive any group of questions since each group is representative of all of the requirements of the license being applied for.

With the test questions, the examinee will receive a separate answer sheet on which he will mark, preferably in pencil, his choice of answer for each numbered question. The use of pencil permits changing an answer if the examinee feels he has made a mistake. The passing grade for the written test is 74% or higher. In a test of 50 questions, 2 points would be taken off for each incorrect answer, allowing no more than 13 wrong answers. The time limit in taking the written test is that of the examination hours scheduled by the FCC office. For example, if the examinations are given from 9 to 11 a.m., there are 2 hours in which to take the examination.

The questions in this text reflect the FCC guide for the Novice, Communicator, General, and Technician class licenses. The answers provided here respond to the specific question asked. However, it is quite possible that on the test form the same question discussed in this text may appear turned around. For example, the question in the text may read, "Which class of amplifier operation is most favorable to the generation of harmonics?" On the examination, the question may be rephrased as, "Which class(es) of amplifier operation do not favor the generation of harmonics?"

Also, a question may appear more complicated than it really is. For example, the problem may be posed by noting in a schematic diagram that the voltage across the resistor in the base circuit of a transistor amplifier measures 4 V. If base current is 1 mA (milliampere), what is the value of the resistor in the base circuit? This is answered by using Ohm's Law, $R = E/I$, for an answer of 4000 Ω (ohms).

Also, the information presented in the text may cover a question that is posed in a manner other than the way it is explained. For example, in a block diagram of a transmitter, the question may ask the examinee to name the circuit represented by the unmarked block in the block diagram rather than ask him to differentiate between the different blocks.

Another possibility is that more than one answer may appear correct. Of the five possible answers, only one is absolutely correct; some of the other answers may be partially correct, but only *one* is absolutely correct.

One way to answer questions that appear to have two or more correct answers is to use the process of elimination. Answers that appear completely false should be ignored. Those answers that appear correct should be studied carefully, and from them answers which appear to be partially correct should be eliminated. Having decided upon the one correct answer, the examinee should compare it with the similar but incorrect answer(s) and try to determine what fact or factors made him decide upon the correct answer. This process should reinforce his reason for selecting the correct answer.

The above discussion should indicate to the reader that rote learning of the question and answer material is not enough. The discussion accompanying the questions and answers in the text also should be studied.

The tests that follow are written expressly to cover Elements 2(A) and (B) and 3(A) and (B), for Novice, Communicator, General, and Technician class licenses; the information for the General class license also is valid for the present Conditional licenses.

Chapter 2

NOVICE CLASS LICENSE: QUESTIONS, ANSWERS, AND DISCUSSIONS

The purpose of the Novice and Communicator class licenses is to encourage more people to become amateur radio operators. It is an ideal way to start since both the code and technical requirements are comparatively simple. For Novice class, the code speed test requirement is 5 wpm, Element 1(A). Basic theory and knowledge of the FCC Rules and Regulations, Element 2(A), are the written test requirements. For Communicator class, however, there is no code requirement, since the Communicator licensee will only operate FM at high frequencies (144-148 MHz). The theory test, Element 2(B), will differ from that of the Novice in that it will contain questions pertaining to high frequency and frequency modulation. An applicant studying for a Communicator license should carefully note those questions pertaining to high frequency and frequency modulation in this chapter; it would also be wise to study the questions on these subjects in Chapter 3, in order to broaden one's background.

The idea behind these two classes of license is to allow a new amateur operator to get on the air with a minimum of effort so that he can enjoy the thrill of communicating with other amateur operators. At present, one must be a citizen of the United States and may not have held an amateur license of any kind for 12 months prior to filing an application for the Novice test, which may only be taken by mail, under the supervision of a volunteer examiner. The Novice license is presently valid for 2 years and may not be renewed. After restructuring, both Novice and Communicator licenses will be taken by mail, although the prerequisites for applicants and requirements for volunteer examiners may change (see Notice of Proposed Rule Making in Appendix III). Also, both licenses will be effective for 5 years and renewable.

The questions, answers, and discussions that follow cover Element 2(A) and 2(B), which are written tests on basic theory and FCC regulations. The questions are based on the latest revised *Study Guide* issued by the FCC and the proposed syllabus-type study guide (Fig. 10). Each question is followed by a short but complete answer similar to those found on FCC tests. When the question refers to the FCC Rules and Regulations, the answer must reflect the Rules and Regulations included in Appendix II, since this is the type of answer required by the FCC. A more detailed discussion follows most answers, in order to provide much needed additional background material for the student. Discussion is omitted on descriptive answers.

The material comprising the study questions for Elements 2(A) and 2(B) is subdivided into the following nine categories:

A – Rules and Regulations
B – Radio Phenomena
C – Operating Procedures
D – Emission Characteristics
E – Electrical Principles
F – Practical Circuits
G – Circuit Components
H – Antennas and Transmission Lines
I – Radio Communication Practices

Each question is labeled by the prefix letter of the category being studied; this is followed by the number 2(A) or 2(B), which indicates Element 2(A) or 2(B), and this, in turn, is followed by the number indicating the order of progression; for example, A.2(A).1, A.2(A).2, B.2(A).1, B.2(A).2, and so on. Questions with two or more parts are identified as (a), (b), etc.; for example, A.2(A).1(a), A.2(A).1(b). For the purposes of this book, since only a small number of the test questions for Elements 2(A) and 2(B) will differ in the type of questions given on each test (as noted above), the questions below have no (A) or (B) in the numbering system.

A. Rules and Regulations

A.2.1 What is the amateur radio service?

Answer: As noted in FCC Rules and Regulations, Section 97.3, *Definitions,* amateur radio service is: "A radio communication service of self-training, intercommunication, and technical investigation carried on by amateur radio operators."

Discussion: The present rules regulating amateur radio service originated in the Communications Act of 1934 in which the Federal Communications Commission was formed. Part 97 of these Rules and Regulations deals with the amateur radio service, and is included in the Appendix II. These Rules and Regulations should be studied and learned, since they are the regulating laws which govern amateur radio operation.

A.2.2(a) What part of the Federal Communications Commission's rules govern the amateur radio service?

Answer: Part 97, *Amateur Radio Service,* governs amateur radio.

A.2.2(b) What are the maximum penalties for violating those rules?

Answer: Upon conviction, the maximum penalty for violating FCC rules is a fine of not more than \$500 for each and every day during which such offense occurs.

Discussion: The penalties for violating the FCC rules are listed under the rules governing commercial radio. (See Section 502 in Appendix II.) Study also Subpart E, *Prohibited Practices and Administrative Sanctions,* Sections 97.112 through 97.137, which point out that the station shall be properly identified, there shall be no remuneration for use of the station, that broadcasting, transmission of music, and deliberate interference are prohibited. This subpart also notes that third-party traffic is prohibited with certain exceptions. Note also that causing interference can lead to restricted operation and that repeated violations can cause license suspension.

A.2.3 What are the fundamental purposes of the amateur radio service?

Answer: To provide voluntary noncommercial communications that contribute to the advancement of the radio art, that encourages improvement of the communication and technical phases of the art, and expands the existing reservoir of trained operators and electronics experts. (See Section 97.1, *Basis and purpose.*)

A.2.4(a) What is the definition of an amateur radio operator?

Answer: A person who holds an FCC amateur radio license and operates solely for personal aim with no pecuniary interest. (See Section 97.3(c).)

A.2.4(b) What is the definition of an amateur radio station?

Answer: A licensed station having the necessary apparatus. (See Section 97.3(e).)

A.2.4(c) What is the definition of a primary station?

Answer: The primary station is the principal amateur radio station shown on the station license. (See Section 97.3 (f).)

A.2.5(a) For how long is a Novice class license valid?

Answer: At present, for 2 years; after restructuring, it will be valid for 5 years.

A.2.5(b) May the Novice class license be renewed?

Answer: At present, a new examination may be taken 1 year after expiration of the Novice license. After restructuring, the Novice license will be renewable. (Refer to Appendix III for the revised Novice class renewal requirements after restructuring.)

Discussion: When a Novice license is retained for the full 2 years, the applicant may apply for another Novice license 1 year after expiration of the license. Novice licensees in the Armed Forces who are sent overseas may ask for a partial renewal (determined by their overseas stay) by writing to the Federal Communications Commission, Washington, D.C. 20554.

A.2.6 May a transmitting station be operated in the Amateur Radio Service without being licensed by the Federal Communications Commission?

Answer: No. (See Section 97.40, *Station license required.*)

A.2.7 Who may hold an amateur radio station license?

Answer: An amateur radio station license will be issued only to a licensed amateur radio operator. Exceptions are made for amateur stations used for military recreation purposes. A license may be issued to a United States citizen in charge of such a station located in approved public quarters. For amateur radio organizations or amateur radio societies, a station license may be issued to an amateur operator (other than Novice or Communicator) as trustee for the organization or society. (See Section 97.37, *General eligibility for station license,* and Section 97.39, *Eligibility of corporations or organizations to hold station license.*)

A.2.8(a) Where must an amateur radio operator license be kept?

Answer: The operator's license must be kept on the operator's person. When operating at a fixed location the license may be posted in a conspicuous place in the room occupied by the operator.

A.2.8(b) Where must an amateur radio station license be kept?

Answer: The station license (or photocopy) must be posted in a conspicuous place in the operating room of a fixed location, or kept on the operator's person. When operating a portable or mobile station, the license (or photocopy) shall be in the personal possession of the person to whom the license has been issued, or a licensed representative.

A.2.8(c) How must an amateur radio station be identified?

Answer: By transmission of the call sign at the beginning and end of each transmission with intervals not to exceed 10 minutes between identification. Portable or mobile operation shall identify by the word "portable" or "mobile" followed by the number of the call sign area in which the station is being operated.

Discussion: Study Section 97.83, *Availability of operator license,* wherein the posting and possession of the operators license is discussed; Section 97.85, *Availability of station license,* which discusses where the station license shall be kept when operating at a fixed or portable location or during mobile operation; and Section 97.87, *Station identification,* which notes how a station shall be identified and at what intervals, as well as how portable or mobile stations shall be identified.

A.2.9 Who is responsible for the proper operation of an amateur station?

Answer: The person to whom the station is licensed (the licensee). (See Section 97.79 (a), *Control operator requirements.*)

A.2.10(a) What is the definition of a control operator?

Answer: The control operator is an amateur operator designated by the station licensee to be responsible for station emissions. (See Section 97.3(p).)

Discussion: A control operator monitors and supervises the radio communication to ensure compliance with the rules.

A.2.10(b) Who may be the control operator of an amateur radio station?

Answer: The control operator may be either the person to whom the station is licensed (licensee) or another amateur radio operator designated by the licensee.

A.2.11(a) What is the log of an amateur radio station?

Answer: The log of an amateur radio station is the record of station transmissions.

A.2.11(b) What information must the log contain?

Answer: The log shall contain the station call sign, signature of the station licensee or a photocopy of the license, the location, and dates of operation.

A.2.11(c) How long should the log be preserved?

Answer: No less than 1 year following the last date of entry.

Discussion: The FCC has relaxed logging requirements. Refer to the Appendix for the latest revision of Section 97.103, **Station log requirements.**

A.2.12 What are the frequency privileges authorized to Novice class licensees?

Answer: 3700–3750 and 7100–7150 kHz; 21.1–21.2 and 28.1–28.2 MHz.

A.2.13 What are the emission privileges authorized to Novice class licensees?

Answer: Type A1 (CW) only.

A.2.14 What is the maximum transmitter power privilege authorized to Novice class licensees?

Answer: The maximum transmitter input power cannot exceed 75 W.

Discussion: Power (*P*), rated in watts (W), can be found by multiplying the voltage (E) in volts (V), and the current (I) in amperes (A) ($P = E \times I$). The input power is the value of plate voltage used in the final amplifier tube of the transmitter multiplied by the current drawn by that tube. For example, a plate voltage of 300 V and a plate current of 0.2 A gives us 300 × 0.2 = 60 W. This value is within the legal limit for input power of a Novice transmitter. After restructuring, refer to Appendix III for the revised Novice class power requirements.

A.2.15 What are the rules regarding the measurement of the frequency of emissions from an amateur radio station?

Answer: The licensee must establish a procedure for measuring the frequency of the transmitter on a periodic basis. This measurement must be made by means independent of the means used to control the frequency generated by the transmitter and must be of sufficient accuracy to assure operation within the frequency band being used. (See Section 97.75, *Frequency measurement and regular check.*)

A.2.16(a) When is one-way communication authorized?

Answer: One-way communication is authorized for emergency communications, drills, information bulletins, round-table discussions, and code practice transmissions.

A.2.16(b) What are "points of communications?"

Answer: Points of communications are the types and locations of radio stations with which amateur radio operators may communicate.

Discussion: Refer to Section 97.89, *Points of communications,* and to Section 97.91, *One-way communications.*

A.2.17 What is the maximum height limitation for an antenna structure?

Answer: The maximum height limitation is 200 feet above ground level.

Discussion: Antenna height limitations are discussed under Section 97.45, *Limitations on antenna structures.* Special limitations for antennas located close to an airport runway are also noted in Section 97.45.

A.2.18 What are the frequencies authorized for the Communicator class licensees?

Answer: All frequencies above 144 MHz.

A.2.19 What are the emission privileges authorized to Communicator class licensees?

Answer: Type F3 (FM) only.

A.2.20 What is the maximum transmitter power privilege authorized to Communicator class licensees?

Answer: The input to the final amplifier stage of the transmitter shall not exceed 250 W.

Discussion: Refer to question A.2.14, above, for a discussion of power. Note that the value of 250 W is the *proposed* power limit in Appendix III; it may be a different value if some modification is made to the proposed new rules prior to adoption.

B. Radio Phenomena

B.2.1(a) How fast do radio waves travel in free space (in meters per second, m/sec)?

Answer: 300,000,000 m/sec.

B.2.1(b) What effect does the medium through which radio waves travel have upon the speed of radio waves?

Answer: The speed of radio waves is reduced when they flow through any medium other than free space (vacuum).

Discussion: Radio waves, the electrical fields created by varying electromagnetic fields that radiate from a transmitting antenna, travel at the speed of light, three hundred million meters per second through free space (a vacuum). Radio waves travel slower through other mediums, with the reduced speed depending upon whether the medium is a good or poor conductor. As the field of radio waves travel outward they spread out to cover larger areas, thus the further away from the source the weaker the signal created by the waves.

B.2.2(a) What is the relationship between the frequency and the wavelength of a radio wave?

Answer: Frequency and wavelength are inversely proportional, i.e., the higher the frequency, the lower the wavelength; the lower the frequency, the higher the wavelength.

B.2.2(b) What are the approximate wavelengths for the frequency bands available to Novice class licensees?

Answer:

Frequency (kHz)	*Wavelength (m)*
3700–3750	80
7100–7150	40
21.1–21.2	15
28.1–28.2	10

Discussion: The formula for finding frequency is:

$$f = \frac{300{,}000{,}000}{\text{wavelength}}$$

where f is measured in hertz; 300,000,000 is measured in meters per second; and the wavelength is measured in meters.

The formula for wavelength is:

$$\text{wavelength} = \frac{300{,}000{,}000}{f}$$

A simpler version is to convert the frequency units to megahertz (MHz) and velocity to 300 so that

$$f\text{ (in MHz)} = \frac{300}{\text{wavelength}} \text{ or wavelength} = \frac{300}{f\text{ (in MHz)}}$$

Referring to Fig. 1, where the Novice frequencies are given in kilohertz, and the corresponding amateur band of frequencies is noted in meters, the reader can see that as the frequency values increase, the wavelength decreases. To verify the wavelength of the band being used, the reader can use as an example the upper edge of the Novice band (frequency equal to 3750 kHz or 3.75 MHz); the procedure is as follows:

$$\text{wavelength} = \frac{300}{f\text{ (MHz)}} = \frac{300}{3.75\text{ MHz}} = 80\text{ m}$$

This shows that the upper edge of the Novice band is exactly at 80 m.

Where the wavelength is known, for example, 40 m, the frequency is found as follows:

$$f\text{ (in MHz)} = \frac{300}{\text{wavelength}} = \frac{300}{40} = 7.5 \text{ MHz}$$

This shows that 40 m itself is out of the Novice band of frequencies, which is 7.10–7.15 MHz.

B.2.3(a) How are radio signals transmitted across great distances?

Answer: Radio signals travel outward from the antenna both along the ground (ground waves) and toward the sky (sky waves) as illustrated in Fig. 11. When the wave travels over the surface of the ground it is referred to as a surface wave. Signals radiated from a vertical antenna, such as a "whip" antenna in a car, are vertically polarized; signals from horizontal antennas, such as a half-wave dipole, are horizontally polarized.

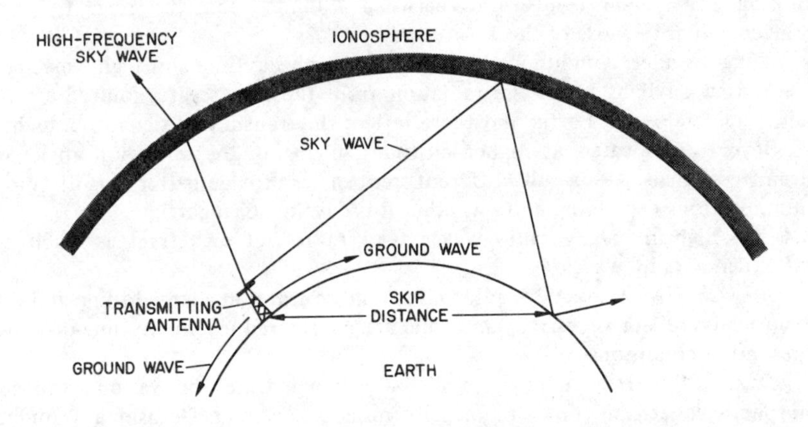

Fig. 11. Radio signal paths.

High-frequency signals that travel out toward the sky strike layers of ionized air (ionosphere) that reflect (bounce) the signals back to Earth at great distances from the transmitting antenna. The distance between the transmitting antenna and the point at which the sky waves return to Earth is called the skip distance. Depending on operating conditions it is possible to operate over multiple-skip distances to provide long-range operation.

B.2.3(b) Which of the amateur radio frequency bands available to Novice class licensees are most likely to result in long-distance communication during the daylight hours? At night?

Answer: During daylight hours, the 15-m band (21.1–21.2 MHz) and/or the 10-m band (28.1–28.2 MHz) will provide the most reliable long-distance communication. At night, the most reliable band for long-distance reception would be the 80-m band (3700–3750 kHz) and/or the 40-m band (7.1–7.15 MHz).

Discussion: The ionosphere level changes between day and night; thus, the time of day will affect the skip distance. Also, the higher the frequency, the greater the angle at which the radio waves leave the antenna. When the frequency reaches wavelengths less than 6 m, the signal strikes the ionosphere at an angle such that the shorter wavelengths penetrate the ionosphere and are not reflected back. As a result, higher frequencies such as the 2-m band are usually considered "line-of-sight" operation, much as a TV station operates. In typical daytime operation, the 80-m Novice band will operate locally and up to 150 miles from the transmitter; the 40-m Novice band typically provides an operating distance of 200 to 400 miles. The 15-m Novice band can provide distances ranging from 1000 miles to skip distances of 2000 or 3000 miles and more.

The Sun will occasionally have storms on its surface, called "sunspots," that directly affect ionospheric conditions. These sunspots vary in 11-year cycles, called sunspot cycles. When these storms become severe the sunspot will create sufficient ionospheric disturbances and changes as to make radio communications range from difficult to impossible.

The 10-m band (28.1–28.2 MHz) varies. Depending upon atmospheric conditions, the 10-m band varies between a limited local coverage to skip distances similar to those of the 15-m band.

Long-distance communication varies between daytime and nighttime, because during daylight hours the radiation from the Sun creates ionized atmospheric layers close to Earth, and these reflect the transmitted signals. At night, these layers are located at higher altitudes, changing the reflection angle of transmitted signals. As a result, different frequencies provide better long distance communications depending upon whether it is daytime or nighttime.

B.2.4 Do high-frequency radio waves tend to reflect or refract as much as low-frequency radio waves?

Answer: High-frequency radio waves tend to travel in a straight line and do not normally reflect or refract and thus are considered essentially line-of-sight for reception conditions.

Discussion: High-frequency radio waves penetrate the various ionized atmospheric layers and do not normally reflect or refract. Occasionally under certain atmospheric conditions, "tropospheric bending" takes place to provide long-distance communication using high frequencies. Another atmospheric phenomenon takes place when portions of the E layer of atmospheric ionization become unusually dense and reflect high frequencies; this is referred to as "sporadic-E" skip. Refer also to question B.3.2(a) of Chapter 3.

C. Operating Procedures

C.2.1 When transmitted by telegraphy, what is the meaning of each of the following: $\overline{\text{CQ}}$*, DE, K, $\overline{\text{AR}}$, $\overline{\text{SK}}$?

Answer: $\overline{\text{CQ}}$ – Calling any station
DE – From
K – Invitation to transmit (go ahead)
$\overline{\text{AR}}$ – End of message
$\overline{\text{SK}}$ – End of transmission

---

*The bar over the two characters indicate that they are run together as one character during transmission.

C.2.2(a) What is the RST reporting system?

Answer: A method of reporting signal readability (R), strength (S), and tone (T).

Discussion: Readability is rated from 1 to 5, with 1 being unreadable and 5 perfectly readable. Strength is rated from 1 to 9 with 1 being faint, barely perceptible, and 9 extremely strong. Tone is rated from 1 to 9 with 1 being extremely rough and 9 a pure dc note. The RST reporting system is shown below:

Readability

1. Unreadable
2. Barely readable, occasional words distinguished
3. Readable with considerable difficulty
4. Readable with practically no difficulty
5. Perfectly readable

Signal Strength

1. Faint signals, barely perceptible
2. Very weak signals
3. Weak signals
4. Fair signals
5. Fairly good signals
6. Good signals
7. Moderately strong signals
8. Strong signals
9. Extremely strong signals

Tone

1. Extremely rough, hissing note
2. Very rough ac note, no trace of musicality
3. Rough low-pitched ac note, slightly musical
4. Rather rough ac note, moderately musical
5. Musically modulated note
6. Modulated note, slight trace of whistle
7. Near dc note, smooth ripple
8. Good dc note, just a trace of ripple
9. Purest dc note

C.2.2(b) What is the meaning of "RST 579"?

Answer: The signal is: perfectly readable; moderately strong; and a pure dc tone.

C.2.3(a) What are "Q" signals?

Answer: Q-signals are a form of abbreviated code in which three-letter signals, all starting with the identifying Q, have been designated by the International Telecommunication Union to have a specific meaning.

C.2.3(b) What is the meaning of: QRM? QRS? QRU? QRZ? QTH? QSL?

Answer:

QRM – I am being interfered with.
QRS – Send more slowly.
QRU – I have nothing for you.
QRZ – You are being called by _______ on _______ kHz.
QTH – My location is _____________.
QSL – I am acknowledging receipt.

Discussion: A complete set of all Q-signals (see Table 5) will be found in Chapter 4, *Station Operation.* It should be noted here that any Q-signal followed by a question mark ($\overline{\text{IMI}}$) changes the meaning of a Q-signal from a statement to a question. For example, QRM ? (QRM $\overline{\text{IMI}}$) becomes "Are you being interfered with?"

C.2.4(a) In what manner should a transmitting frequency be selected for an amateur radio station?

Answer: The frequency band should be selected for the transmission distance desired and should be of a frequency for which the licensee is eligible. The frequency selected should be clear to prevent interference with other operators.

C.2.4(b) What additional factors should be considered when selecting a transmitting frequency near one end of the authorized frequency band?

Answer: The accuracy of the oscillator being used to set the output frequency of the transmitter should be known to prevent the frequency of transmission from going outside of the authorized frequency used as the band edge.

Discussion: When using a variable frequency oscillator (VFO) to set the output frequency near the edge of a specified band, such as the 3700–3750 kHz, 80-m Novice band, the operator must be careful not to have the output frequency go below or above the band edges. The frequency of an oscillator can be checked with a crystal-controlled, secondary-standard oscillator having a typical error of 0.01%. The error of 0.01% is multiplied by the frequency of the band edge being checked. The answer will tell how close to the upper or lower band edges one can operate using the frequency meter as a monitor. To simplify: multiply, using kilohertz, and convert the 0.01% by dividing 100 by 0.01 to obtain a decimal value of 0.0001 (the answer will be in kilohertz). For example, at the upper edge of the 80-m Novice band, 3750 kHz, multiply 3750 by 0.0001 for a value of 0.3750 kHz. This means that using the frequency meter as a guide to the accuracy of the oscillator, one should operate no closer to the upper edge of the band than 3749.6250 kHz.

The discussion pertains to the accuracy of the frequency meter, and when operating CW, it implies no bandwidth. In practice, a bandwidth of about 100 Hz is considered normal for CW keying at approximately 20 wpm. To be sure operation is not above or below the band edge while working CW, it is prudent to allow a minimum of another 50 Hz to the 0.01% tolerance value.

As the band frequency increases, the value of the 0.01% error also increases. For example, the allowable tolerance at the upper limit of the 80-m Novice band is 0.3750 kHz. For the upper limit of the 40-m Novice band, one has 7150 kHz × 0.0001, or a 0.7150-kHz error tolerance. Thus, the frequency tolerance of the Novice 40-m upper band edge limit is approximately twice the tolerance limit of the upper limit of the 80-m Novice band edge.

C.2.5(a) What is Public Service operating?

Answer: Amateurs who handle noncommercial messages in traffic nets, or work with emergency nets, are considered to be performing a public service.

C.2.5(b) What is message traffic in a network operation?

Answer: A group of amateur radio operators who operate regularly on a specified frequency to handle messages either within a local area or to pass the messages to their destination, form a network. The messages they handle are referred to as message "traffic."

C.2.6(a) On FM bands, what is meant by operating simplex?

Answer: When two stations are transmitting and receiving each other on the same frequency.

C.2.6(b) On FM bands, what is meant by operating duplex?

Answer: When the two stations are working each other through a repeater using one frequency for transmitting and a different frequency for reception.

Discussion: The 144–148 MHz band has certain specific frequencies commonly agreed upon for direct operation between stations; this type of operation is called "simplex." When working another station through a repeater (discussed below under category *F. Practical Circuits*), transmission is on one specific frequency and the repeater re-transmits the signal on another specific frequency to which the receivers are tuned.

D. Emission Characteristics

D.2.1 What are A0 and A1 emissions?

Answer: A0 emission is the emission of an unmodulated carrier. A1 emission is radiotelegraphy by the on-off keying of a transmitter whose output is an unmodulated signal.

Discussion: The description of types A0 and A1 is as follows:

Type of Modulation	*Type of Transmission*	*Symbol*
Amplitude	No modulation (carrier only)	A0
Amplitude	Telegraphy without the use of modulating audio frequency (by on-off keying).	A1
	Telegraphy by the on-off keying of an amplitude modulating audio frequency or audio frequencies or by the on-off keying of the modulated emission (special case: an unkeyed emission amplitude modulated).	

Telegraphy by on-off keying without the use of modulating audio frequency is used by amateur radio operators. It consists of an oscillator, whose continuous wave signal (carrier) is keyed on and off to transmit code. The phrase "without the use of modulating audio frequency" refers to the type of transmitter in which the oscillator signal is modulated by an audio frequency. The modulation is keyed on and off, not the transmitter oscillator. The two types of output signals are illustrated in Fig. 12. Each type of A1 transmission requires a specific type of detector at the radio receiver.

D.2.2 What are the characteristics of a good quality A1 emission?

Answer: The transmitted signal has no spurious radiation, such as that caused by key clicks, chirp, or backwave, and it should be a pure dc note.

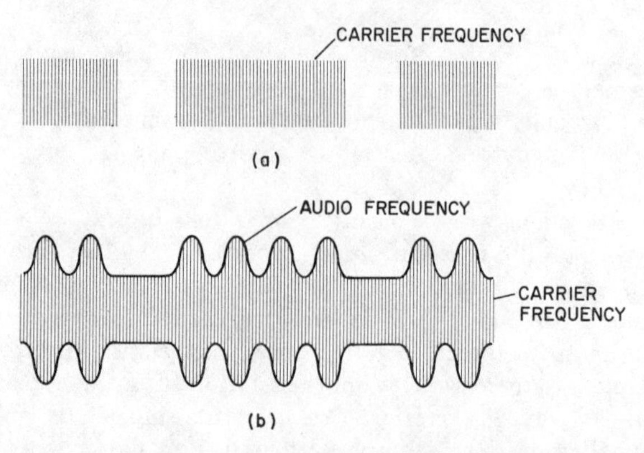

Fig. 12. Type A1 keying of the letter R. (a) Only the keyed carrier signal is transmitted. (b) Using an on-off keyed audio modulation signal superimposed on the carrier.

Discussion: Spurious radiation is the generation of frequencies on either side of the transmitted frequency, which create interference with other signals. Key clicks are caused by the sudden opening and closing of the key, which creates an arc. The arc is removed by a filter in the keying circuit. Chirp is a shift in the oscillator output frequency caused by the load on the oscillator circuit created by keying the circuit. Chirp is removed by oscillator isolation and regulation of the voltage source. Backwave is the "leaking" of the oscillator signal from the transmitter when the key is up, which causes undesired signal radiation. Backwave is removed by proper shielding and by isolating the oscillator and using special keying circuits.

D.2.3 What method(s) of keying provides maximum frequency stability?

Answer: Keying an amplifier stage.

Discussion: Keying the oscillator usually results in the oscillator shifting frequency to some degree. The best method of keying a transmitter is to key an amplifier stage following the oscillator.

D.2.4 Describe amplitude modulation.

Answer: In amplitude modulation, the transmitter carrier frequency is fixed and the amplitude (height or level) of the signal is varied (modulated) by the speech signal.

D.2.5(a) What is type A3 emission?

Answer: Type A3 emission is amplitude-modulated radiotelephony.

D.2.5(b) What type of emission is single sideband (SSB)?

Answer: Single-sideband emission is type A3.

Discussion: Refer to questions D.3.1 through D.3.4 in Chapter 3 for additional information on amplitude modulation.

D.2.6 Describe frequency modulation.

Answer: In frequency modulation, the transmitter carrier amplitude is fixed and the carrier frequency is varied (modulated) by the speech signal.

D.2.7(a) What is type F3 emission?

Answer: Type F3 emission is frequency-modulated radiotelephony.

D.2.7(b) What is wideband F3 emission?

Answer: Wideband F3 emission is frequency-modulated radiotelephony in which the sidebands generated by modulating signals are no greater than 15 kHz on either side of the carrier frequency, for a total maximum bandwidth of 30 kHz.

D.2.7(c) What is narrowband F3 emission?

Answer: Narrowband F3 emission is frequency-modulated radiotelephony in which the sidebands generated by modulating signals are no greater than 3 kHz on either side of the carrier frequency, for a total maximum bandwidth of 6 kHz.

D.2.7(d) What type of F3 bandwidth is permitted in the 144–148 MHz band?

Answer: Wideband F3 (30-kHz bandwidth) is permitted in the 144–148 MHz band.

Discussion: Refer to questions D.3.1 and D.3.3 through D.3.6 in Chapter 3 for additional information on frequency modulation.

E. Electrical Principles

E.2.1(a) What is electromotive force?

Answer: Electromotive force (EMF), more commonly called voltage, is a form of electrical energy generated by a difference in potential. This difference in potential causes current to flow in a complete (closed) circuit.

E.2.1(b) What is current?

Answer: Current is a flow of electrons through a closed circuit. The flow is created by electromotive force applied to the circuit.

E.2.1(c) What is electrical power?

Answer: Electrical power is the rate at which electromotive force, forcing current through a closed circuit, will do work.

E.2.1(d) What units of measurement are used for electromotive force, current, electrical power?

Answer: Electromotive force–volts; current–amperes; electrical power–watts.

Discussion: Electromotive force provides the force that propels electrons through a circuit. The unit for EMF, the volt, is used to measure the difference in potential. The "potential" refers to the surplus of electrons at the negative connection of the source (a cell is a classic example of a source) and shortage of electrons at the positive connection. The source, or cell, has the potential (ability) to provide electromotive force. A typical basic cell will have a voltage value of about 1.5 V.

When a closed circuit is applied across the terminals of the source of potential, electrons flow from the negative terminal through the resistance offered by the circuit to the positive terminal. Current is measured by the number of electrons that pass a given point in a specific time. Approximately 6300 trillion (specifically 6.28×10^{18}) electrons must travel past one point in 1 sec to equal 1 A.

Power, the rate of doing work, is equal to the value of the voltage multiplied by the current: Power (P) = Voltage (E) $\times$ Current (I), or, P(watts) = E(volts) $\times$ I(amperes). Note that the symbol for voltage or electromotive force is E, and that for current it is I, representing Intensity. When the resistance of the circuit is known, the value of resistance can be used with the value of voltage, or current, to determine the power being expended. Either of the following formulas (the symbol for resistance is R) may be used: $P = E^2/R$ or $P = I^2 \times R$.

E.2.2(a) What is direct current?

Answer: Direct current is where the flow of electrons in a closed circuit always moves in the same direction.

E.2.2(b) What is alternating current?

Answer: Alternating current is current flow caused by a voltage source that constantly changes its polarity. Because of the periodic reversing of voltage polarity, the current flows through the circuit first in one direction, then in the opposite direction. The number of times per second that the voltage polarity reverses is the *frequency* of the alternating current.

E.2.2(c) How can alternating current be converted to direct current?

Answer: By applying the alternating current to a rectifier circuit. The rectifier (usually a diode) will allow the flow of electrons in one direction only to change alternating current to pulsating direct current.

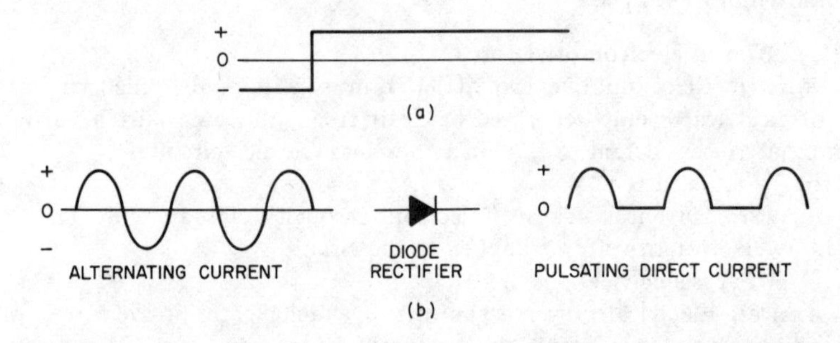

Fig. 13. (a) Graphic illustration of direct current. (b) Graphic illustration of alternating current and rectification of alternating current to direct current by use of a diode rectifier.

Discussion: Batteries are excellent sources of direct current (Fig. 13a) particularly where portability is essential or where equipment is being used in areas without conventional power. Alternating current (Fig. 13b), which is used for power in homes, is derived from generators designed to provide an output voltage at the typical home outlet socket of 115–120 V. The voltage polarities change at a rate of 60 times per second. Since most equipment circuits operate with dc, it is convenient to convert the ac from the house wiring to dc to power the circuits. Figure 13(b) shows the alternating current being converted to pulsating direct current when passed through a rectifier.

E.2.3(a) What is a series circuit?

Answer: When current flows in a continuous path through two or more components we have a "series" circuit.

E.2.3(b) What is a parallel circuit?

Answer: When current divides between two or more components or circuits we have a "parallel" circuit.

Discussion: As shown in Fig. 14(a), in a series circuit current flows through each resistor. Since the current is continuous through each resistor the value of current is equal throughout the circuit. In Fig. 14(b) the current divides through each parallel branch circuit. Since each parallel branch circuit will have

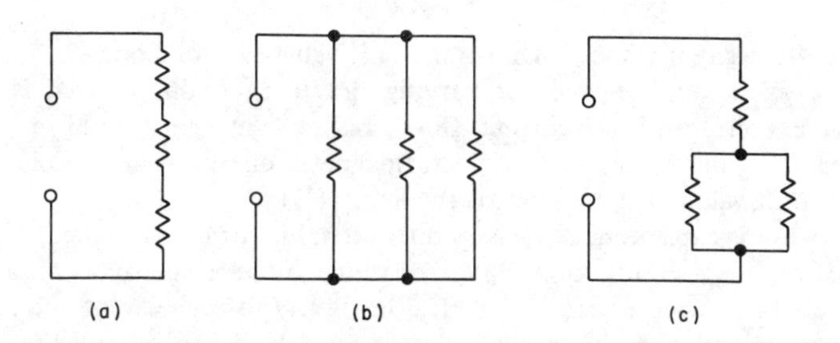

Fig. 14. Basic circuits: (a) series; (b) parallel; (c) series-parallel.

the amount of current flow determined by the resistance of that branch, the current flow can differ in each branch. In Fig. 14(c) we have a series-parallel circuit in which the total current will flow through the series resistor and the total current will then divide between the two parallel resistors.

E.2.4 What is a cycle? A kilocycle? A megacycle? A hertz? A kilohertz? A megahertz?

Answer: A cycle is one complete change in 1 sec. A kilocycle equals 1000 cycles; a megacycle equals 1,000,000 cycles.

A hertz is 1 cps (cycle per second), used for measuring frequency. A kilohertz equals 1000 Hz; a megahertz equals 1,000,000 Hz.

Discussion: As noted in Chapter 1, the term hertz has replaced the term cycle per second. A cycle per second (1 cps) equals a hertz (1 Hz); a kilocycle (1 kc) equals a kilohertz (1 kHz); and a megacycle (1 Mc) equals a megahertz (1 MHz).

E.2.5 What is rf?

Answer: Radio frequencies (rf) are those frequencies above 15,000 Hz (15 kHz). Signals at these frequencies can radiate over a distance to transmit a radio signal.

Discussion: Frequencies from 15 to 20,000 Hz are called audio frequencies (af). Frequencies greater than 15,000 Hz are capable of radiating a signal as noted above. Signals at the overlapping frequencies of 15,000 to 20,000 Hz are considered very poor radiating signals and are not generally used.

E.2.6 What is the relationship between a fundamental frequency and its second harmonic? Its third harmonic?

Answer: A harmonic frequency is a whole number multiple (*integral*) of the original or fundamental frequency. A *second* harmonic is two times the frequency of the fundamental; a *third* harmonic, three times; and so on.

Discussion: The frequency to which a circuit is tuned is called the fundamental frequency. Harmonic frequencies are directly related to the fundamental frequency by an integral multiple, i.e., 2, 3, 4, 5 times, and so on. Note that there is *no* first harmonic since the first harmonic is the fundamental frequency itself. The fundamental frequency times one (the first harmonic) would simply be the fundamental frequency itself.

E.2.7(a) What is resistance? What is the unit value for resistance?

Answer: Resistance is the opposition of a material to the flow of current. The unit value of resistance is the *ohm* (Ω).

E.2.7(b) What is inductance? What is the unit value for inductance?

Answer: In applying ac (or varying dc) to a conductor that is usually wound in a coil form the current flow creates a magnetic field around the conductor. This magnetic field is electromagnetic energy that is called inductance. The unit value of inductance is the henry (H).

E.2.7(c) What is capacitance? What is the unit value for capacitance?

Answer: Two conducting plates separated by an insulator form a capacitor, which can be used to store electrical energy. The capacity of the plates to accept positive and negative charges determines the value of capacitance. The unit value of capacitance is the farad (F).

Discussion: Any device that is a poor conductor and is used in a circuit specifically to offer opposition to current flow is called a resistor. Resistors (Fig. 15a) can be formed from carbon compositions (carbon resistor) or from special wire that offers a high resistance to the current flow (wire-wound resistor). Since the value of resistance in ohms is low, the values of resistance are most often given in thousands of ohms (kilohms) or millions of ohms (megohms).

Inductance theoretically takes place in any length of wire carrying ac. However, the value of inductance is related to the frequency of the alternating current, and at low and middle frequencies, a short length of wire offers so little inductive value that it can be ignored. Inductance may be increased by increasing the magnetic field. Wrapping the wire in a spiral coil (Fig. 15b) so the magnetic fields of each turn of wire overlap provides a stronger total field, thus increasing the value of inductance. The symbol for an inductor is shown in Fig. 15c. The unit value of inductance, the henry, is rather large. For convenience, the units used are thousandths of a henry (millihenry) or millionths of a henry (microhenry).

Capacitance can vary greatly in value, type (fixed or variable), and operating voltage. Conductor plates with small surface area offer little capacitance. Specially etched foil plates rolled into tubes increase the surface area which, in turn, increases the capacitance value. Spacing and the type of insulator used to separate the two conducting storage plates also affect the capacitance value. Most capacitors are fixed in value and can operate with no higher operating voltage than that determined by the type of insulator used between the capacitor plates. Some capacitors have one set of fixed plates against which another set of movable plates can be adjusted to create more or less capacity. These are called variable capacitors. The symbol for a fixed capacitor is shown in Fig. 15d, those for a variable capacitor in Fig. 15e.

The unit value of capacitance, the farad, is very large. Most often capacitance is rated in millionths of a farad (microfarad).

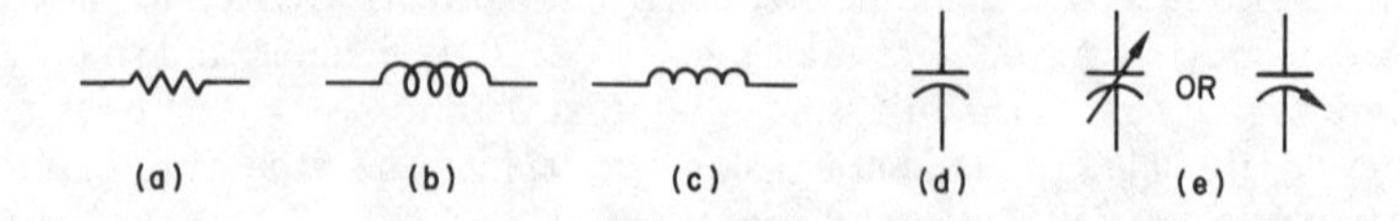

Fig. 15. (a) Resistor symbol. (b) Wrapping a wire in a coil shape to increase inductance. (c) Inductor symbol. (d) Fixed capacitor symbol. (e) Variable capacitor symbol.

Unit values of resistance (ohms) are indicated by the symbol Greek letter omega (Ω). Unit values of inductance (henries) are indicated by the letter H. Unit values of capacitance (farads) are indicated by the letter F. Kilo (k) is used to express thousands. For example, two thousand ohms would be written 2 kΩ; two million ohms would be 2 MΩ. The value for a thousandth of a unit is milli (m). A millionth of a unit is a micro (μ). Two millihenries would be written 2 mH, and two microfarads, 2 μF.

F. Practical Circuits

F.2.1 Draw the schematic diagram of a circuit having the following components: (a) Battery with internal resistance; (b) resistive load; (c) voltmeter; (d) ammeter.

Answer:

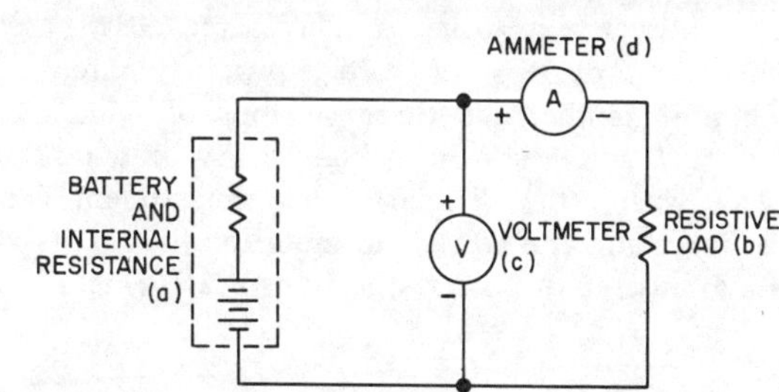

Discussion: There is no fixed method of indicating internal resistance in a battery. The most common method is to draw the symbol for a battery in series with a resistor and enclose both with a dashed outline to indicate that they are combined. Note that the voltmeter is across the source and the current meter is in series with the source.

F.2.2 Using the circuit shown in F.2.1, how can the value of the resistor load be determined? How can the power consumed by the load be determined?

Answer: The value of the resistive load in the circuit above is determined by applying Ohm's Law, $R = E/I$. The power consumed by the load can be determined by using the following equation, $P = E \times I$. The values of E(voltage) and I (amperes) are shown on the meters.

Discussion: Ohm's Law expresses the direct relation between the voltage, current, and resistance in a circuit. It is an algebraic equation stated as

$$I \text{ (current in amperes)} = \frac{E \text{ (voltage in volts)}}{R \text{ (resistance in ohms)}}$$

Algebraically this can be restated as $R = E/I$, and $E = I \times R$. Thus, with any two values known, the third value can be found.

The relationship expressed by Ohm's Law is very important. Changing one value will cause a fixed (linear) change of another value. For example, in a circuit containing 100 Ω resistance and measuring 200 V, E/R = 200 V/100 Ω = 2 A of current flowing in the circuit. If the voltage is doubled to 400 V, 400 V/100 Ω = 4 A, or double the original value of current flow. If, instead, the voltage is left at 200 V and one-half of the original resistance is used, 50 Ω, 200 V/50 Ω, will give a value of 4 A of current flow.

Leaving the resistance fixed and doubling the voltage doubles the current; leaving the voltage fixed and halving the resistance also doubles the current. Thus, current varies directly with voltage and inversely with resistance.

To find the resistance of a circuit having 100 V and 1 A of current,

$$R = \frac{E}{I} = \frac{100\text{ V}}{1\text{ A}} = 100\ \Omega$$

If the current flowing through the circuit were equal to 2 A, then the resistance would be halved (R = 100 V/2 A = 50 Ω). If the current remained at 1 A and only 50 V were applied, again the resistance would be 50 Ω (R = 50 V/1 A).

To find the voltage given the current and resistance, $E = I \times R$. Here again the relationship is direct. Increasing or decreasing either the current or the resistance will increase or decrease the voltage proportionately.

To help him remember the three versions of Ohm's Law, the student should learn the *Pie Diagram* shown in Fig. 16. By drawing the three values in their proper place within the divided circle, the student can cover any one element with a finger and the correct formula for finding its value will remain uncovered. For a discussion of power, questions E.2.1(c) and E.2.1(d).

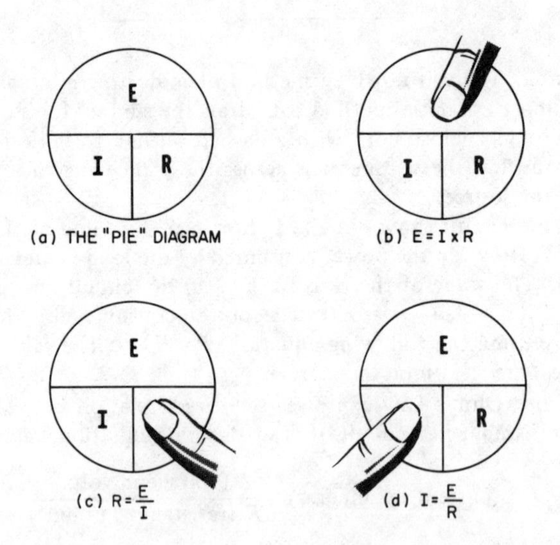

Fig. 16. Using the "Pie Diagram" to find the three versions of Ohm's Law.

F.2.3 In the circuit shown in F.2.1, what must the value of the resistive load be in order for the maximum power to be delivered from the battery?

Answer: The value of the resistive load should be equal to (match) the value of the internal resistance of the battery.

Discussion: Maximum power is delivered to a load when the resistance of the load matches the resistance of the source. For example, a battery has a

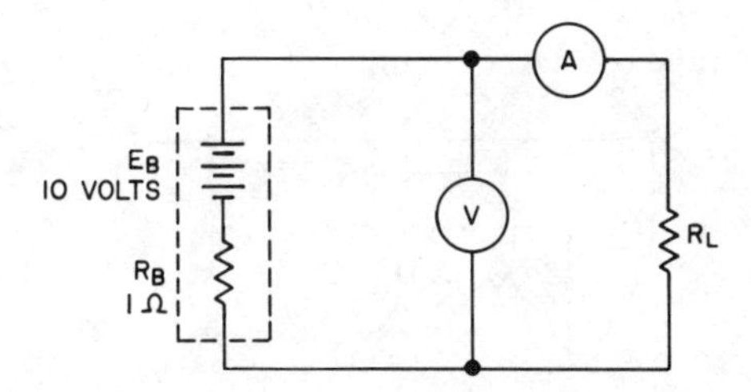

Fig. 17. Maximum power output is achieved when the value of the resistive load matches the internal resistance value of the source.

potential of 10 V and an internal resistance (R_B) of 1 Ω, as shown in Fig. 17. To illustrate the value of matching, the load resistors (R_L) are given values, 0.33 Ω, 0.5 Ω, 1 Ω, 2 Ω, and 3 Ω. Placing the 0.33 Ω resistor as R_L the total circuit resistance ($R_T = R_B + R_L$) is 1 + 0.33 = 1.33 Ω. Using Ohm's Law, $I = E/R$, $I =$ 10 V/1.33 Ω, I = 7.5 A. This value of 7.5 A is the same throughout the circuit; that is, 7.5 A flow through both R_B and R_L. The next step is to determine the voltage drops across R_B and R_L using the circuit current value of 7.5 A. To do this, use $E = I \times R$. The voltage drop across R_B is $E = 7.5 \times 1 = 7.5$. The voltage drop across R_L is E = 7.5 × 0.33 = 2.475, rounded off to 2.5 V. The final step is to determine the power dissipated in R_L, using the equation $P = E \times I$ (refer to E.2.1). The value of E across R_L is 2.5 V; therefore, P (in watts) of R_L = 2.5 V × 7.5 A = 18.75 W.

Table 5 lists the circuit values obtained with each value of resistive load, showing that maximum power is delivered to the load when the resistive value of the load matches the internal resistance of the source.

Table 5. Circuit Values of Resistive Loads.

E_B	R_B	R_L	R_T	I	E of R_B	E of R_L	P of R_L
10 V	1 Ω	0.33Ω	1.33Ω	7.5A	7.5V	2.5V	18.75W
10 V	1 Ω	0.5Ω	1.5Ω	6.67A	6.67V	3.33V	22.2W
10 V	1 Ω	1Ω	2Ω	5A	5V	5V	25W
10 V	1 Ω	2Ω	3Ω	3.33A	3.33V	6.67V	22.2W
10 V	1 Ω	3Ω	4Ω	2.5A	2.5V	7.5V	18.75W

F.2.4 Draw a basic tube amplifier circuit.

Answer:

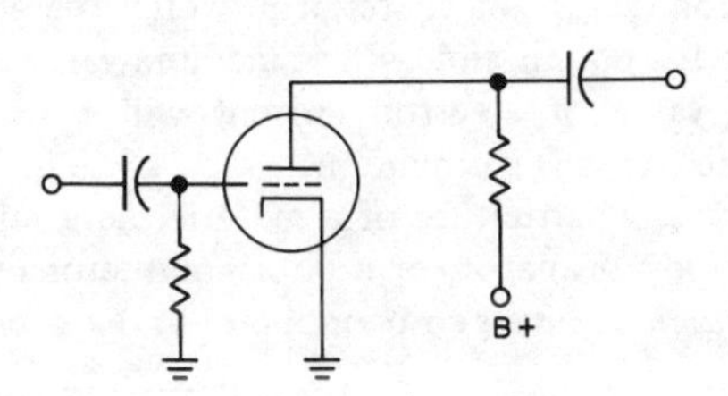

F.2.5 Draw a basic tube oscillator circuit.

Answer:

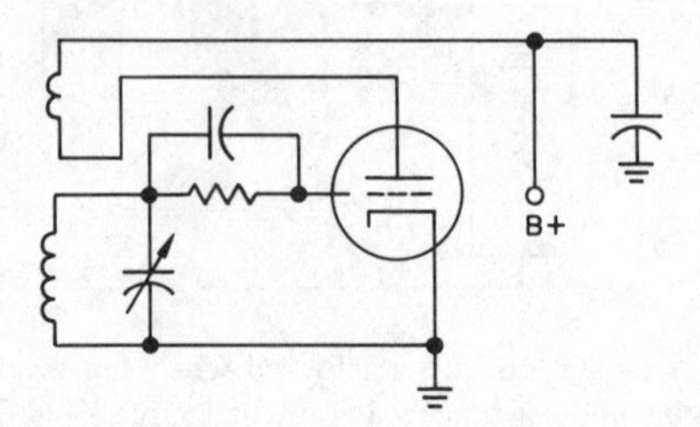

F.2.6 Draw a transformer-operated half-wave rectifier circuit and capacitor-input filter. Use solid-state diodes.

Answer:

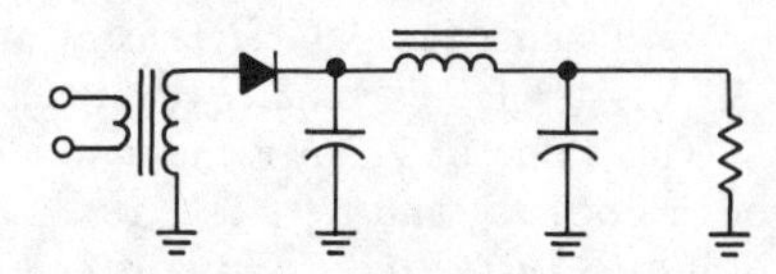

F.2.7 Draw a transformer-operated full-wave rectifier circuit and choke-input filter. Use solid-state diodes.

Answer:

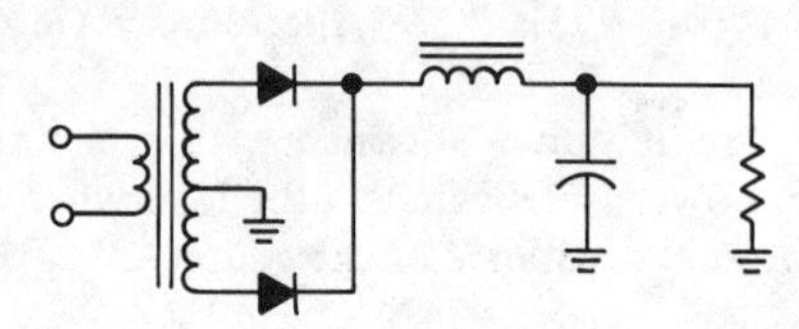

G. Circuit Components

G.2.1(a) What is an insulator? Give an example.

Answer: An insulator is a material in which the atomic electrons are tightly bound and, even with a high value of electrical potential applied, no appreciable electron current will flow. An example of an insulator is glass.

G.2.1(b) What is a conductor? Give an example.

Answer: A conductor is a material in which the atomic electrons are loosely bound and, even with a relatively low value of electrical potential applied, a high value of electron current will flow. An example of a conductor is copper wire.

G.2.1(c) What is a semiconductor? Give an example.

Answer: A semiconductor is a material in which the atomic electrons are neither loosely nor tightly bound and, with a medium value of electric potential applied, a measurable value of electron current will flow. An example of a semiconductor is impure (doped) germanium.

Discussion: The atomic structure of a material determines whether it is a good conductor or a good insulator, or a poor conductor or poor insulator. A material that falls between being a good conductor and a good insulator is called a semiconductor. Since most materials fall either into the category of insulator or conductor the best-known semiconductor is man-made. In their pure states germanium and silicon are insulators, but by adding impurities (doping) they become semiconductors and are widely used in solid-state diodes and transistors.

G.2.1(d) What is a crystal?

Answer: A crystal is the name given to certain crystalline materials that will develop a voltage when pressure is applied to the surfaces, or they will vibrate when a signal is applied to the surfaces.

Discussion: Certain crystalline materials, of which Rochelle salts is the best known natural crystal material, will develop a voltage when a narrow slab of the material is placed between two conducting plates and the crystal is squeezed. Conversely, if an electrical signal is applied to the plates the crystal will vibrate at the rate of the signal. This action is known as the "piezoelectric" effect.

G.2.2 Draw the schematic symbol for a resistor, a capacitor, an inductor, a transformer, a choke, and a crystal.

Answer:

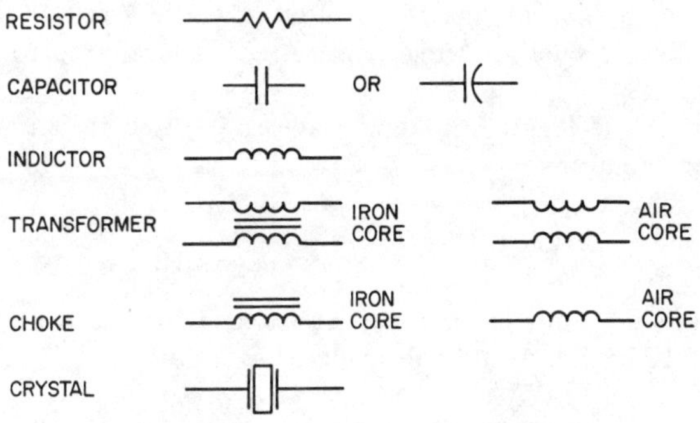

RESISTOR
CAPACITOR OR
INDUCTOR
TRANSFORMER IRON CORE AIR CORE
CHOKE IRON CORE AIR CORE
CRYSTAL

Discussion: The symbol for an inductor, a coil of wire, is usually drawn as shown above. An inductor using an iron core is referred to as a choke. The iron core is indicated by parallel lines drawn above the inductor symbol. A transformer consists of two or more wire coils. The coils may be wound one on the other or may be placed next to each other. The air core transformer is drawn as two inductors opposite each other. For transformers using an iron core, the iron core is indicated by parallel lines drawn between the two inductors. Either an air core or iron core transformer symbol is correct; however, the majority of transformers are iron core types, and its symbol is better known.

G.2.3 Draw the schematic symbol of a diode; a transistor; a triode vacuum tube; a tetrode vacuum tube; a pentode vacuum tube.

Answer:

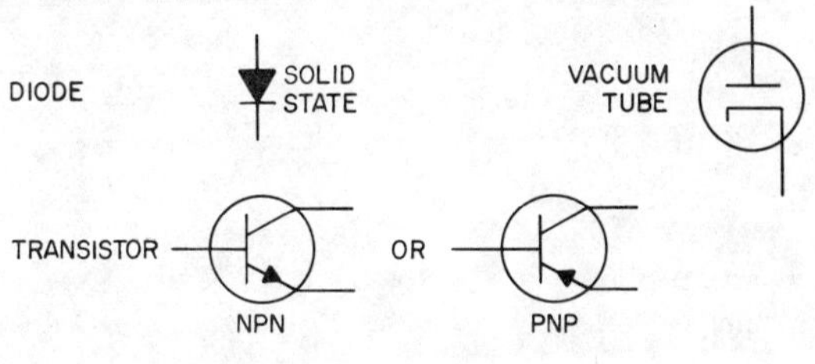

DIODE SOLID STATE VACUUM TUBE
TRANSISTOR OR
NPN PNP

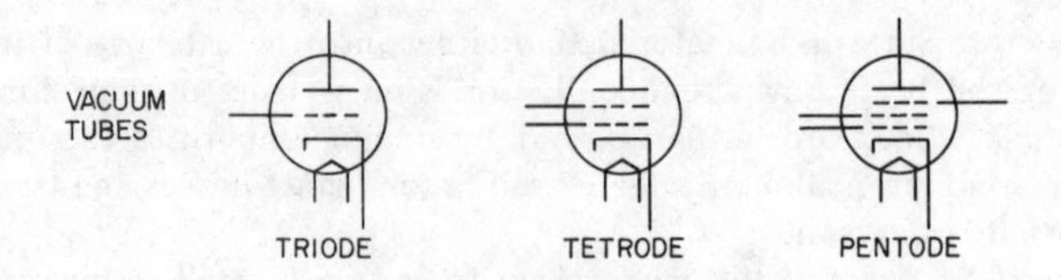

Discussion: The symbol for a diode may be either that shown for the solid-state type or that of a vacuum tube type. The symbol for a transistor may be for one of many types, junction transistor, FET, and the like; however, the most common type of transistor is either the NPN or PNP shown above. Operation of diodes, transistors, and tubes are explained in question G.3.1 of Chapter 3.

H. Antennas and Transmission Lines

H.2.1 What is a dipole antenna?

Answer: A dipole antenna is an antenna whose length is one-half of the wavelength of the frequency being transmitted, also referred to as a half-wave antenna.

Discussion: The length of a half wave to be used for an antenna can be found using the equation,

$$\text{length (feet)} = \frac{468}{\text{frequency (MHz)}}$$

The length of a dipole antenna cut (tuned) for the center of the 40-m Novice band, 7.125 MHz, would be calculated as follows:

$$\frac{468}{7.125} = 65.6 \text{ feet, rounded to 65 feet 6 inches}$$

The normal equation for finding the length in feet of a half-wave antenna is:

$$\text{length (feet)} = \frac{468}{\text{frequency (MHz)}}$$

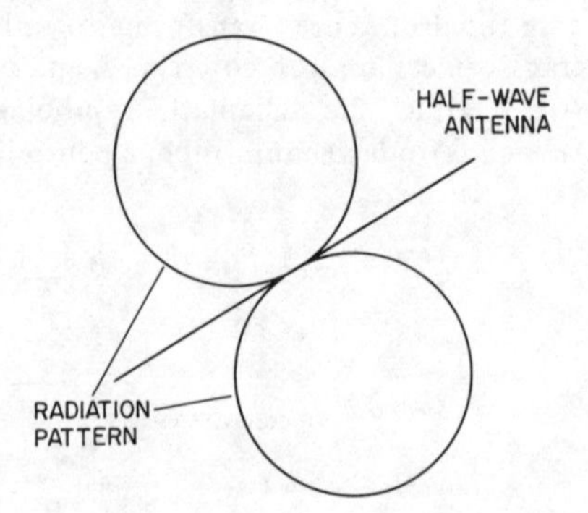

Fig. 18. Radiation pattern of a half-wave dipole antenna.

However, in actual use an antenna exhibits what is called *end effect.* To overcome the end effect, the antenna is shortened by a factor that changes the 492 to 468. Because the end of the antenna offers a very high impedance in the order of several thousand ohms, whereas the center of the antenna offers an impedance of approximately 70 Ω, the antenna is usually operated center fed with a coaxial cable (discussed in H.2.3, below). The strength of the signal (field strength) radiated by the dipole antenna creates a radiated pattern that is predominantly off the sides of the antenna (see Fig. 18).

H.2.2(a) What is a half-wave antenna?

Answer: A half-wave antenna is an antenna whose length is one-half the wavelength of the frequency being transmitted.

H.2.2(b) What are the approximate lengths (in feet) for half-wave antennas for the frequency bands authorized for Novice class licensees?

Answer: Using the formula noted in H.2.1,

$$\text{length (feet)} = \frac{468}{\text{frequency (MHz)}}$$

and selecting the center of the Novice bands yields the following:

$$\text{80-m band (3.725 MHz):} \quad \frac{468}{3.725} = 125.6 \text{ feet}$$

$$\text{40-m band (7.125 MHz):} \quad \frac{468}{7.125} = 65.6 \text{ feet}$$

$$\text{15-m band (21.15 MHz):} \quad \frac{468}{21.15} = 22.1 \text{ feet}$$

$$\text{10-m band (28.15 MHz):} \quad \frac{468}{28.15} = 16.6 \text{ feet}$$

Discussion: A half-wave antenna can be used at harmonic frequencies such as two times the original frequency (second harmonic) or three times (third harmonic) but the impedance at the center of the antenna will vary with the application of the higher frequencies.

H.2.3(a) What is a transmission line?

Answer: A transmission line is used to carry the radio-frequency energy from the transmitter to the antenna.

Discussion: Ideally, transmission lines should "match" the output impedance of the transmitter and the input impedance of the antenna. In addition, the transmission line should have low losses.

Impedance is the value of the combined dc and ac resistances in a circuit. As noted in question F.2.3, maximum power is delivered to a load (the antenna) when the resistive load matches the value of the internal resistance of the source (the transmitter). However, the antenna is usually located some distance from the transmitter. By using a transmission line whose impedance matches that of the antenna (72 Ω for a typical dipole), there will be no loss at the transmission line-to-antenna connection. If the transmission line impedance does not match the antenna impedance, the mismatch develops voltage and current waves along the transmission line, called *standing waves.* (Refer to question H.3.4(a) and (b) in Chapter 3, for a detailed discussion of standing waves.) By connecting the transmission line at the transmitter to an output circuit such as the pi-network, the pi-network can be tuned to match the impedance of the transmission line. In this manner, the maximum output can be produced at the antenna, since there is a match from the transmitter output to the antenna.

H.2.3(b) What are some commonly used transmission lines?

Answer: Single wire, parallel conductor, and coaxial lines.

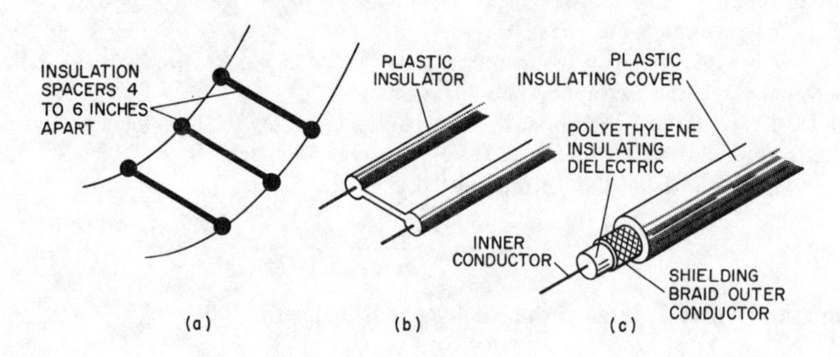

Fig. 19. Transmission line types: (a) open wire; (b) ribbon (twin lead); (c) coaxial cable.

Discussion: As shown in Fig. 19(a) and (b), parallel conductor transmission lines are either open-wire or plastic-covered. The open-wire type can use different width spacers, which will vary the impedance of the line, and the wider the spacer the higher the impedance. Typical impedance values for open-wire transmission lines are between 400 and 600 Ω. The plastic-covered transmission line, also known as the ribbon type or *twin-lead,* is commonly used for TV antenna installations. It has an impedance of 300 Ω and, although inexpensive, it has higher losses than that of the open-wire type.

Another and more commonly used type of transmission line is the coaxial cable shown in Fig. 19(c). It has a center conductor in a circular polyethylene insulator surrounded by a woven wire braid which is covered by insulating plastic; the braid is used as the second or ground conductor. Coaxial cable impedance varies with the thickness of the center insulating material, but the majority of coaxial cables offer 50- or 72-Ω impedance.

H.2.4(a) What are some advantages of a multiband antenna?

Answer: Convenience is the primary advantage of a multiband antenna. One antenna will cover two or more bands of frequencies, thus removing the need for switching circuits between antennas. It also requires less space and is less expensive than other antennas.

H.2.4(b) What are some disadvantages of a multiband antenna?

Answer: The multiband antenna will work efficiently only at the band to which its physical dimensions are closest. Impedance matching devices, which provide increased losses but prevent radiation of harmonic frequencies, are usually required to have the antenna and transmission line match the output of the transmitter at the different bands of frequencies.

H.2.5(a) What is meant by antenna polarization?

Answer: The signal radiated from a transmitting antenna will follow the polarization (position) of the antenna. A vertical antenna transmits in a vertical plane and a horizontal antenna in a horizontal plane.

H.2.5(b) For FM transmission, what is the most used type of polarization?

Answer: Vertical.

Discussion: Close to the transmitting antenna the polarization of the signal is determined by the antenna position. At a distance, upon reception, the polarization of high-frequency signals is usually not clearly vertical or horizontal but somewhere in between. It is considered good practice to have the receiving antenna polarized in the same pattern as the transmitting antenna. Since the majority of repeater stations and mobile transmitters use vertical polarization, FM transmission and reception is predominantly by vertically polarized antennas. Refer to question B.3.4(a) of Chapter 3.

I. Radio Communication Practices

I.2.1 What precautions can be taken to reduce the possibility of shock hazard in amateur radio stations?

Answer: The major precautions are:

1. The equipment should be enclosed to prevent accidental contact with exposed high voltage points.
2. All metal chassis should be connected directly to ground, preferably to a metal water pipe at the inlet to the building or to an external ground rod. If this is not practical, the chassis should return to electrical ground through an approved grounded conductor as noted in the following precaution.
3. The wall outlet used to supply ac voltage should be an Underwriters Laboratories approved 3-prong grounded outlet; the power cable from the equipment to the wall outlet should be a matching 2-wire plus groundwire (3-conductor) cable.
4. Power supplies should use bleeder resistors across the filter output to discharge the capacitors of the filter circuit when the power supply is turned off.
5. External feed lines to the antenna, and the antenna itself, should be placed at a sufficient distance from any power lines to prevent their contact with the power lines should the antenna, or the power lines, be blown or knocked down.

I.2.2 How can radio or television interference by amateur radio transmission be prevented?

Answer: Be sure that the transmitter has no spurious radiation due to generation of harmonics or parasitic oscillations.

Discussion: Improper tuning of a transmitter can cause generation of harmonic frequencies. A self-oscillating (parasitic oscillator) circuit in the transmitter can also develop undesired radiating frequencies. These harmonic and parasitic oscillation signals can create interference with radio (radio frequency interference, RFI) or television (television interference, TVI) receivers.

I.2.3(a) Illustrate the correct method of connecting a voltmeter and ammeter.

Answer:

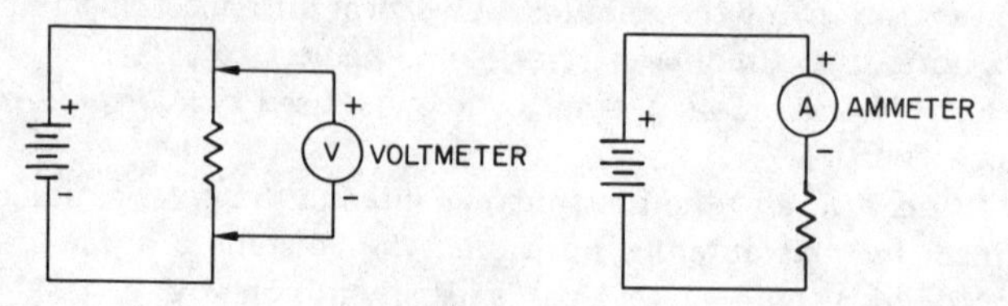

Discussion: The voltmeter is connected parallel to the voltage source to be measured with the polarity of the meter matching the polarity of the source. The ammeter is connected in series with the current flow of the circuit being measured with the positive terminal of the meter connected toward the positive terminal of the source.

I.2.3(b) What precaution(s) must be taken when using an ohmmeter?

Answer: There should be no current flowing in the circuit under test and the circuit should be checked to make sure that there is no parallel or shunting circuit across the circuit being checked.

I.2.4 What is the power input to a vacuum tube in the final amplifier stage of a transmitter, exclusive of power for heating the cathode, for the following operating conditions:

Driving power	–	0.5 W
Plate voltage	–	600 V
Plate current	–	140 mA
Screen voltage	–	175 V
Screen current	–	10 mA
Filament voltage	–	6.3 V
Filament current	–	0.8 A

Answer: $P = E \times I$, $P = 600 \times 0.14$, $P = 84$ W. Add to this the 0.5 W of driving power to provide a total input power of 84.5 W. (Note that this exceeds the 75-W limit permitted a Novice station.)

Discussion: Power input is the dc power input of the plate circuit of the vacuum tube. It is determined by multiplying the plate voltage by the plate current ($P = E \times I$). For the set of conditions given above, the values of plate voltage and plate current are the only ones required to derive the value of power input for the vacuum tube in the final amplifier. Power supplied from a preceding driver stage must be included as part of the total input power rating.

To convert milliamperes to amperes in computing the answer above, divide the 140 mA by 1000 to give 0.14 A.

I.2.5 What methods do amateur radio licensees most often use to determine whether an emission from a transmitter is within an authorized frequency band?

Answer: The methods available include a calibrated wavemeter, a calibrated receiver, or a frequency meter. Calibration can be achieved by comparing the wavemeter, frequency meter, or receiver with the standard-frequency broadcast by the National Bureau of Standards stations WWV, WWVH, WWVB, etc.

Discussion: Bureau of Standards stations transmit (with high accuracy) signals on the following frequencies: 2.5, 5, 10, 15, 20, and 25 MHz. These signals are sufficiently accurate to allow calibration of the wavemeter, frequency meter, or receiver. The availability and the relative simplicity of calibrating receivers account for their frequent use in determining the transmitter output frequency.

I.2.6 What methods do amateur radio licensees most frequently use to determine the quality of emissions from their station?

Answer: The most common method is to use the signal report given by the station being worked by the operator. Another method is to use the station receiver to check the transmitter output.

Discussion: To check the transmitter output using the receiver, the transmitter should be connected to a dummy load in place of the antenna so that test signals are not radiated, which causes interference with other operators on the air. The antenna lead to the receiver should be disconnected to prevent overloading the receiver. While transmitting to the dummy load, the signal should be checked with the receiver for emission quality, key clicks, chirp, and the dc note. The receiver should be tuned to either side of the transmitting frequency to check for quality of emission and for spurious signals.

I.2.7(a) What is a transmatch?

Answer: A transmatch is a special coupling circuit used between the output of the transmitter and the transmission line to have the output impedance of the transmitter match the transmission line impedance.

I.2.7(b) What are the advantages of using a transmatch?

Answer: The transmatch permits coupling transmitter output circuits to unmatched transmission lines. It provides a matching output to the transmitter for maximum power output to the antenna and reduces the power of harmonic frequencies to help prevent undesired harmonic radiation.

Discussion: A transmatch used as shown in Fig. 20 will provide an impedance match between the transmission line and the output circuit of the transmitter. This is particularly useful if the transmission line to be used is not the correct impedance to match that of the transmitter output circuit. For example, a transmatch can be used where a transmitter output circuit provides a low impedance unbalanced (one side grounded) output for use with coaxial cable, but where the transmission line being used is a high-impedance parallel conductor type. (Refer to question F.2.3 for matching, and to H.2.3 for transmission line information.)

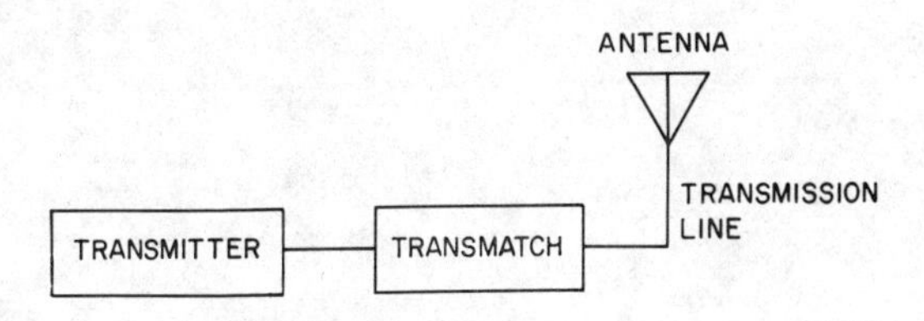

Fig. 20. Using a transmatch to have the transmission line impedance match the transmitter output impedance.

I.2.8 What is a repeater?

Answer: A repeater is a receiver-transmitter combination in which a signal is received at one frequency and re-transmitted at another frequency on the same band.

Discussion: A repeater is used to provide increased communications range resulting from the limited line-of-sight operating range usually found using high-frequency FM transmissions. The repeater is placed at a high elevation so that most if not all transmissions will be received because of its advantageous location. The repeater automatically re-transmits, on another frequency, any received signal. Since the repeater is usually at a high elevation, its transmitted signal will be available over a significantly greater area than that available on a line-of-sight basis between either mobile (automobile) or fixed (home) FM receivers and transmitters.

Repeaters are usually automatic, self-operated, and unattended. Special licenses are required to operate repeaters, and the repeater station has a special call sign. Refer to Sections 97.3, 97.40, 97.41, 97.67, 97.89, 97.109, 97.110, and 97.111 of the *FCC Rules and Regulations* in Appendix II.

Chapter 3

GENERAL CLASS LICENSE: QUESTIONS, ANSWERS, AND DISCUSSIONS

Technician, Conditional, and General Licenses

Technician License

The Technician class license is expressly designed to encourage experimentation and development of the higher frequency amateur bands. Technician license requirements include the same test in theory and FCC regulations as that given for the General class license, but the code speed is the same as that for a Novice, 5 wpm. The Technician license permits operation of all types: AM and FM, CW, MCW, phone, TV, or facsimile. Operation is permitted only on specified very-high-frequency bands, and on some bands crystal must be used for frequency control.

The Technician license is presently applied for by mail. All steps are identical to those of the Novice license application, including the code test requirements. The only exceptions are that Form 610 is filled out for a Technician class license and a filing fee is required when applying for the written examination. If the applicant for a Technician license holds a Novice license and passes the Technician test, his Novice license is canceled and he can use only the privileges of the Technician license. A Technician class license must be used only with the privileges accorded to that class of license. The license is good for 5 years and is continually renewable.

After restructuring, an amateur who elects to operate with a Series B license (frequencies above 29 MHz) will first become a Communicator by passing the test for Element 2(B). A Communicator who desires to become a Technician must then take Element 1(A) by mail under the supervision of volunteer examiners; the next step is to make an appointment at an FCC office to take the examination for Element 3(B). If an amateur desires to start directly as a Technician he may take both Elements 1(A) and 2(B) together under the supervision of volunteer examiners (these elements will be given only by mail) prior to taking Element 3(B) at an FCC office. Element 3(B), the written test for the Technician license, differs from Element 3(A), the test for the General class license, only in that it will contain special questions on high-frequency operation.

Conditional License

The Conditional license is presently identical to the General license, the only difference being that the applicant is unable to take the test in person at an FCC engineering office or field office. The applicant for a Conditional license requests Form 610 from the nearest FCC field engineering office. The form is filled out for a Conditional class license, and a qualified volunteer examiner is located. A filing fee is required when the application for the written examination is submitted.

A Conditional class license must be used with the privileges accorded to that class of license. The license is good for 5 years and continually renewable. If the Conditional class licensee should move to an area that is within the 175-mile distance limit, he need not reapply for a General class license but can renew the Conditional class license indefinitely.

After restructuring, there will be no Conditional class license per se. Current holders of Conditional class licenses may convert them to either the Technician (C) or General (C) licenses, or both (refer to Table 1 in Appendix III). The designation (C) is a temporary conditional authorization allowing operation until the licensee can appear before an FCC examiner to qualify in person. These licenses will *not* be renewable since the FCC feels that a contionally licensed operator who is serious about his craft should be able to appear at an FCC office *sometime* within the 5-year period of the license (see Notice #17 of Proposed Rule Making in Appendix III).

General License

The General class license is perhaps the best-known license because it is held by the majority of amateur operators. As shown in Fig. 1, General class licensees can presently operate in all but small portions of the frequency bands set aside for Advanced and Extra class licensees. General class licensees have no restrictions. They can operate with any authorized type of emission: A1, continuous-wave telegraphy; A3, amplitude modulation (AM) or single sideband (SSB); A5, television (TV); and so on.

Code speed test requirements are those of Element 1(B), 13 wpm; and the applicant must pass a written test on Element 3, general amateur practice and FCC regulations. The test for a General Class license can be taken only at an FCC office. The license is effective for 5 years.

The questions used in this text are based upon the latest revised *Study Guide* issued by the FCC and the proposed syllabus-type study guide. An FCC-type syllabus for the General class is shown in Fig. 21. Each question is followed by a short but complete answer, such as that found in the FCC test. When the question refers to FCC Rules and Regulations the answer must, of necessity, reflect the Rules and Regulations included in Appendix II, since this is the type of answer required by the FCC. A more detailed discussion follows the answer to provide additional background material. The discussion is omitted for descriptive answers.

Fig. 21. FCC-type Syllabus Study Guide for Element 3: General class amateur radio operator's license.

A. Rules and Regulations

A.3.1 *Definitions:* Amateur radio; Amateur radio service; Third-party traffic; Control point; Remote control; Control station; Control operator; Emergency communications; Fixed operation; Portable operation; Mobile operation.
A.3.2 *Operator privileges:* Frequencies; Emissions; Transmitter power.
A.3.3 *Station operation:* Control stations; Model control; Remotely controlled stations.
A.3.4 *Responsibilities:* Volunteer examiner.
A.3.5 *Prohibited practices:* Remuneration; Music.

B. Radio Phenomena

B.3.1 *Definitions:* Maximum useable frequency; D-layer, E-layer, Sporadic-E layer; Critical frequency; F-layer.
B.3.2 *Wave propagation:* Ionospheric propagation in the HF and VHF amateur bands: Effects of sunspot activity.

C. Operating Procedures

C.3.1 *Radio interference:* Responsibility; Elimination or reduction.
C.3.2 *Harmonics and parasitics:* Generation; Reduction; Avoidance.

D. Emission Characteristics

D.3.1 *Definitions:* Peak-envelope power; Frequency deviation; S/D ratio; Occupied bandwidth; Sidebands.
D.3.2 *Classification of emissions:* A2; A3; F0; F3.
D.3.3 *Modulation:* Characteristics and types; conveying of voice intelligence; Voice intelligibility range; Percentage of modulation.
D.3.4 *Power relations in speech waveforms:* Determination of peak-envelope power; Determination of average power; Determination of the peak-to-average power ratio.
D.3.5 *General factors concerning emissions:* Types of distortion: amplitude, frequency, phase; Causes, elimination, or reduction of distortion.

E. Electrical Principles

E.3.1 *Definitions:* Impedance; Reactance; Decibels; Mutual inductance; Circuit Q; Resonance.
E.3.2 *Circuit theory:* Ohm's law; Capacitors in series and parallel; Inductors in series and parallel; Voltage division across capacitors; Voltage division across inductors; Inductive reactance; Capacitive reactance; Series resonance; Parallel resonance; Determination of resonant frequency; Circuit Q; Impedance matching.
E.3.3 *Principles of transformers:* Operating principles; Transformer efficiency; Turns ratio.
E.3.4 *Principles of magnetism:* Induced voltage.

Fig. 21. FCC-type Syllabus Study Guide for Element 3: General class amateur radio operator's license (*continued*).

F. Practical Circuits

F.3.1 *Definitions:* Secondary emission; Transit time; Interelectrode capacitance.
F.3.2 *General circuits:* Oscillator operation; Oscillator loading; Grid bias; Cathode bias; Grounded-grid, grounded-plate, grounded-cathode circuits.
F.3.3 *Filter circuits:* Low-pass; High-pass; Bandpass; Band reject.
F.3.4 *Rectifier circuits:* Peak inverse voltage; Series operation of diodes.
F.3.5 *Power supplies:* Voltage regulation.
F.3.6 *Coupling circuits:* Direct coupling; Capacitive coupling; Impedance coupling; Link coupling.
F.3.7 *Transmitter circuits:* Amplitude modulated transmission; Frequency modulated transmission; Power ratings; Filter method SSB; Phasing method SSB.
F.3.8 *Receiver circuits:* Beat frequency oscillator; Automatic gain control.

G. Circuit Components

G.3.1 *Components:* Electrolytic capacitors; Toroidal inductors.
G.3.2 *Solid-state components:* Diodes; Transistors.
G.3.3 *Vacuum tubes:* Characteristics; Plate current saturation; Amplification factor; Phase inversion; Characteristic curves and load lines.
G.3.4 *Crystals:* Piezoelectric effect; Thickness versus frequency.

H. Antennas and Transmission Lines

H.3.1 *Definitions:* Radiation resistance; Parasitic excitation; End effect; Antenna resonance; Antenna impedance.
H.3.2 *Types of antennas:* Multielement arrays; Grounded antennas; Half-wave dipoles.
H.3.3 *Antenna characteristics:* Polarization; Directivity; Effective power gain.
H.3.4 *Transmission lines:* Characteristic impedance; Radiation losses; Attenuation versus frequency.

I. Radio Communication Practices

I.3.1 *Functional diagrams:* SSB transmitters; FM transmitter; AM transmitter.
I.3.2 *Tests, measurements, and adjustments:* Percentage modulation tests; Two-tone tests; Linear amplifier adjustments; Distortion tests.
I.3.3 *Test equipment:* Oscilloscope; Calibrated receiver; Marker generator; Reflectometer.
I.3.4 *Receiver principles:* Intermodulation; Sensitivity; Selectivity; Desensitization; Thermal, man-made, and atmospheric noise.
I.3.5 *Transmitter principles:* Cross-modulation.

The material comprising the study questions for Elements 3(A) and 3(B) is subdivided into the following nine categories:

A – Rules and Regulations
B – Radio Phenomena
C – Operating Procedures
D – Emission Characteristics
E – Electrical Principles
F – Practical Circuits
G – Circuit Components
H – Antennas and Transmission Lines
I – Radio Communication Practices

Each question is labeled by the prefix letter of the category being studied; this is followed by the number 3(A) or 3(B), which indicates Element 3(A) or 3(B), and this, in turn, is followed by the number indicating the order or progression; for example, A.3(A).1, A.3(A).2, B.3(A).1, B.3(A).2, and so on. Questions with two or more parts are identified as (a), (b), (c), etc.; for example, A.3(a).1(a), A.3(A).1(b). For the purposes of this book, since only a small number of the test questions for Elements 3(A) and 3(B) will differ in the type of questions given on each test (as noted above), the questions below have no (A) or (B) in the numbering system.

A. Rules and Regulations

A.3.1 What are the five principles expressing the fundamental purpose of the Amateur Radio Service?

Answer: The five principles are:

1. A voluntary noncommercial communication service, particularly with respect to providing emergency communications;
2. The amateur's ability to contribute to the advancement of the radio art;
3. Advancing skills in both the communication and technical phases of the art;
4. A reservoir of trained operators, technicians, and electronic experts;
5. The amateurs ability to enhance international good will.

(See Appendix II, *FCC Rules and Regulations,* Part 97, *Amateur Radio Service,* Section 97.1, *Basis and purpose.*)

A.3.2 What is the definition of amateur radio communication?

Answer: Noncommercial radio communication between amateur radio stations for personal aim only, not for business or money interests. (See Section 97.3(b), *Definitions.*)

A.3.3 What is the definition of fixed operation? Portable operation? Mobile operation?

Answer: Fixed operation is from the specific location shown on the station license. Portable operation is from a specific location other than that shown on the station license. Mobile operation is while in motion or during a stop at an unspecified location. (See Section 97.3(m).)

A.3.4 With what stations may an amateur radio station communicate?

Answer: An amateur radio station may communicate with any other radio amateur station except those of countries whose administration objects to such radiocommunications; any FCC licensed radio station operating for Civil Defense purposes, or any station approved by the FCC to communicate with amateurs. (See Section 97.89, *Points of Communication,* and Subpart H, Appendix 2, Article 41, Sec. 1.)

A.3.5 What types of transmissions may be made by United States amateur radio stations to amateur radio stations in foreign countries?

Answer: Where permitted, transmissions shall be made in plain language and shall be limited to messages of a personal or of a technical nature. Transmitting a communication on behalf of a third party is absolutely forbidden unless a special arrangement has been made between the United States and the foreign country.

Discussion: An amateur may not handle a message on behalf of a party other than the foreign amateur operator with whom he is in radio communication. This ruling is intended to prevent any unauthorized use of amateur radio communications to evade the public telecommunications systems of foreign countries.

These regulations are noted in Subpart H of the International Regulations (listed in the Appendix) as Appendix 2, Article 41, Sec. 2. Refer also to Section 97.3(w) for additional limitations on third-party traffic.

A.3.6 What is the definition of third-party traffic?

Answer: A radio communication between amateur operators that is on behalf of anyone other than the control operators of the amateur stations. (See Section 97.3(w).)

Discussion: Third-party agreements are not common and those countries that do permit it usually make various stipulations; such as, the types of messages handled for a third party must be such that they could not be sent by other normal channels of radio communications. It is best to check with the FCC (or radio amateur journals) for details as to the type of third-party message handling currently permitted.

As of this writing the countries having third-party agreements with the United States are:

Argentina	Costa Rica	Haiti	Panama
Bolivia	Cuba	Honduras	Paraguay
Brazil	Dominican Republic	Israel	Peru
Canada	Ecuador	Liberia	Trinidad and Tobago
Chile	El Salvador	Mexico	Uruguay
Colombia	Guyana	Nicaragua	Venezuela

A.3.7 Under what limitations may a third party participate in amateur radio communications?

Answer: A third party may participate in amateur radio communication provided that a licensed radio amateur is acting as the control operator. (See Section 97.79(d).)

A.3.8(a) What types of one-way transmissions by amateur radio stations are permitted?

Answer: The following types of one-way transmissions to amateur radio stations are authorized: emergency communications including bona fide emergency practice drills, amateur radio information bulletins, net (round-table) communications, and code practice transmissions. (See Sections 97.89, *Points of Communication,* and 97.91, *One-way communications.*)

A.3.8(b) What types of one-way transmissions by amateur radio stations are prohibited?

Answer: Amateur radio stations are prohibited from public broadcasting and transmission of music. Transmissions for which the amateur operator receives remuneration.

Discussion: There are no prohibited practices specifically relating to one-way transmissions. The types of one-way transmissions that are prohibited are listed as broadcasting and transmitting music; these are referred to in Sections 97.113 and 97.115.

A.3.9 May an amateur or radio station automatically retransmit programs or signals emanating from any class of station other than amateur?

Answer: No.

Discussion: As noted above, amateur radio shall not broadcast, which includes "the retransmission by automatic means of programs or signals emanating from any class of station other than amateur."

A.3.10 When there is a violation of the rules at an amateur radio station, who is held responsible?

Answer: The station licensee is always responsible. (See Section 97.79, *Control operator requirements.*)

A.3.11(a) What is the minimum information that must be recorded in a log for every amateur radio station?

Answer: The log shall contain the station call sign, signature of the station licensee or a photocopy of the license, the location, and dates of operation.

A.3.11(b) How long must the log be preserved?

Answer: Not less than 1 year following the last date of entry.

Discussion: The FCC has relaxed its logging requirements. Refer to Appendix II for the latest revision of Section 97.103, *Station log requirements.*

A.3.12 What determines the operator privileges at an amateur radio station where the control operator is other than the station licensee? How should the station be identified?

Answer: The operator privileges are determined by the class of license held by the control operator.

The control operator should use the call sign of the station from which he is operating. When the control operator has a license whose privileges exceed those of the station license, he may operate in the privileged frequencies permitted the control operator, but he must follow the station call sign with the call sign of the control operator.

Discussion: Control operator requirements spell out the answer to such questions as "When an operator with a General class license operates a Novice class station what call should he use?" If the operation of the Novice station follows all the Novice restrictions, the station call sign is that of the Novice station license. If operation is to be on frequencies assigned to holders of General class licenses, then the call sign of the General class licensee will follow that of the Novice licensee. For example; CQ DE WN1XYZ/W1ABC. (See Sections 97.79(c) and 97.87(d).)

A.3.13(a) What is the definition of a control point?

Answer: The control point is the operating position of an amateur radio station where the control operator function is performed. (See Section 97.3(q).)

A.3.13(b) Which types of amateur stations must have a control point?

Answer: Every type of amateur station must have at least one control point. (See Section 97.43, *Location of station.*)

A.3.13(c) What is the definition of a control station.

Answer: A control station is the station licensed to conduct remote control of another amateur radio station. (See Section 97.3 (i), *Additional station.*)

A.3.14 In what manner, and at what intervals, must an amateur radio station be identified by the transmission of its call sign?

Answer: By transmitting the station call sign at the beginning and end of each transmission, and at intervals not to exceed 10 minutes during any one or more transmissions. (See Section 97.87, *Station identification.*)

A.3.15 What are the rules regarding the transmission of interference? Music? Codes and Ciphers? Obscenity? Indecency? Profanity? False signals? Unidentified radio communications or signals?

Answer: The transmission of interference, music, codes and ciphers, obscenity, indecency, profanity, false signals, and unidentified radio communications or signals are expressly prohibited. Upon conviction, the maximum penalty for violation of these rules is a fine of not more than $500 for each and every day during which such offense occurs.

Discussion: The complete listing of prohibited practices is noted in Appendix II, Part 97, Subpart E, under *Prohibited Transmissions and Practices,* Sections 97.112 through 97.129. The penalties for violating FCC rules are listed in Appendix II under *Excerpts from the Communications Act, 1934,* Section 502.

A.3.16(a) When is a notice of operation away from the authorized location of an amateur radio station required?

Answer: Notice is required when operating portable or when operating at a fixed location away from the authorized location for more than 15 days.

A.3.16(b) Where must the notice (of operation away from the authorized location of an amateur radio station) be sent?

Answer: Advance written notice must be given to the Engineer in Charge of the FCC radio district in which operation is intended.

Discussion: Although the question asks only when and where the notice should be sent if operating away from the authorized location, the following information is pertinent to portable operation. Previously, portable operation for more than 48 hours required notification. The change to 15 days became effective in October 1972. A 15-day operation permits operating portable on a 2-week vacation without having to give notice to the local FCC district in which the vacation is being spent. However, operating in a district other than the one in which the station license has been obtained requires that the call sign be given, followed first by the word "portable" and then by the number of the call sign of the area in which the station is being operated. (See Section 97.87(b)(2).) For example, if W2ABC is vacationing in Massachusetts, the call would be "W2ABC portable 1."

Notice is required for vacation periods exceeding 15 days, or for students attending school, etc. The Rules and Regulations that must be followed for portable operation or operation at a fixed location for an extended period of time include: Section 97.97, *Notice of operation away from authorized location;* Section 97.95, *Operation away from the authorized permanent station location;* and Section 97.87, *Station identification.*

A.3.17 What are the requirements in order to qualify for the special provisions in the rules for stations used only for radio control of remote craft and vehicles?

Answer: The requirements are that the mean power output not exceed 1 W and that the amateur transmitter must have affixed to it an FCC card (FCC Form 452-C) or a durable plate indicating the station call sign and licensee's name and address.

Discussion: As noted in Appendix II, Subpart D, under *Special Provisions,* Section 97.99 provides for operation of remote control model aircraft and vehicles by an amateur station.

A.3.18(a) What are the consequences, should the holder of an amateur radio operator license obtained by mail examination under the supervision of a volunteer examiner, be ordered to appear for a Commission-supervised examination and fail to do so, or fail to pass the examination?

Answer: The operator license is subject to cancellation, and a new license will not be issued for the same class operator license as that canceled. (See Section 97.35(a).

A.3.18(b) What are the responsibilities of a volunteer examiner?

Answer: The volunteer examiner is responsible for the proper conduct and supervision of the examination.

Discussion: The volunteer examiner must give the examination in strict accordance with the instructions provided by the FCC. (Refer to Section 97.29(b)(2).)

A.3.19 What is the maximum authorized power for the holder of a radio amateur license other than Novice?

Answer: Maximum authorized power is an input power of 1 kW to the plate circuit of the final amplifier stage.

Discussion: Present maximum authorized power, as noted in Section 97.67, allows a maximum input power of 1 kW. Under the proposed restructuring, authorized maximum power leves are scheduled to change to the following: Novice and Communicator, 250 W input power; Technician and General, 500 W input power; Experimenter, Advanced, and Extra, 2 kW PEP (peak envelope power). Refer to the proposed amendments to Sections 97.7 and 97.67 in Appendix III. These proposed values may change when the restructured regulations, with any final modifications, are passed. Be *sure* to check the permitted power levels allowed by the FCC after restructuring takes effect.

B. Radio Phenomena

B.3.1 What are the propagation characteristics of the HF and VHF amateur frequency bands?

Answer: The HF band will provide skip distance operation with the skip distances varying with the frequencies selected and the time of day. The VHF band is essentially local operation over line-of-sight distances.

Discussion: The HF band ranges from 3–30 MHz, the VHF band from 30–300 MHz. The HF band has the following bands of amateur frequencies; 1.8 MHz (160 m), 3.5 MHz (80 m), 7 MHz (40 m), 14 MHz (20 m), 21 MHz (15 m), 28 MHz (10 m). In the VHF band the more commonly used bands of amateur frequencies are 50 MHz (6 m) and 144 MHz (2 m).

The 160-m band operates mainly on local ground waves during the day and at night can provide skip transmission for approximately 1000 miles or more.

The 80-m band provides short skip distances of 100–200 miles in the day and at night has approximately the same propagation characteristics as the 160-m band.

The 40-m band has typical daytime skip distances of 500–800 miles which increase at night to distances of 2000 miles or more.

The 20-m band has longer daytime skip distance, typically 1000–2000 miles. During a sunspot cycle, the daytime range can be increased to about twice the normal range. Operation is usually sporadic at night and is good only during a sunspot cycle.

The 15-m band varies greatly with the sunspot cycle. During the day, propagation characteristics are similar to those of the 20-m band. During sunspot cycles, the band can also be used at night for long-distance transmission.

The 10-m band is generally dormant and usable only for local reception. At the peak of a sunspot cycle, the band will be useful for approximately the same distances and during the same hours as those of the 15-m band.

In the 6-m, 2-m, and higher UHF bands, operation is basically line-of-sight; occasional weather changes can create "ducts" that provide long-distance operation.

Occasionally, disturbances on the Sun's surface creates atmospheric disturbance. These surface disturbances, called "sunspots," vary in time and intensity. The number of sunspots peaks in 11-year cycles. During a sunspot disturbance, the number and height of ionized layers create atmospheric changes (ionospheric storms). Since skip distance is determined by the presence and height of these ionized layers, changes in the atmosphere caused by sunspots will change propagation conditions.

B.3.2(a) What are some propagation factors that influence radio transmission and reception on the amateur frequency bands?

Answer: The presence or lack of ionized bands in the atmosphere, sunspot activity, and the signal frequency all influence amateur radio transmission and reception.

Discussion: A major factor affecting transmission and reception on the amateur frequency bands is the presence or lack of ionized bands in the Earth's atmosphere. Ionization results from the action of the ultraviolet rays, gamma

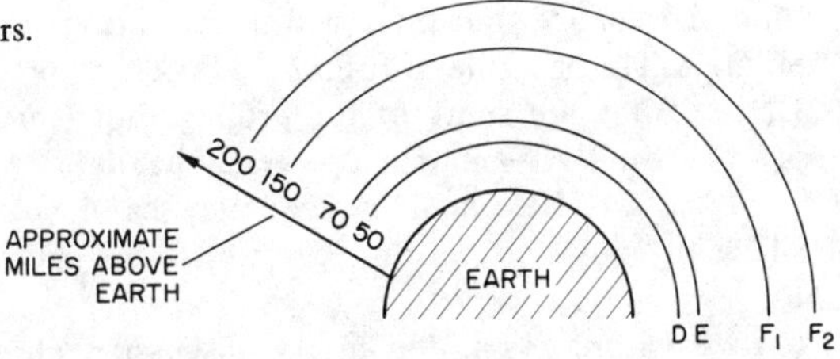

Fig. 22. Daytime ionization layers.

rays, and other particles emitted by the Sun. Because they are a product of the Sun's rays, the ionized bands change with the time of day. During daytime, there are four bands, the D, E, F1, and F2. Their approximate distances above Earth are 50, 70, 150, and 200 miles, respectively (Fig. 22). At night, the D and E layers disappear and the F1 and F2 layers merge to form a single F layer at approximately 175 miles above the Earth.

Sunspot activity causes the ionized layers to form at distances higher than normal, thus increasing the skip distance. Figure 23 illustrates this principle. A signal leaving Earth at a fixed angle will reflect back to Earth at a further distance when sunspot activity has caused the ionized layer to rise.

The frequency of the signal determines what interaction will occur when the signal strikes the ionized layer. As shown in Figs. 11 and 22, the signal strikes the ionized layer, is bent, and returns to Earth. The bending of the signal in the ionized layer is called refraction. At lower frequencies (longer wavelengths), the refraction is more pronounced; thus, the lower frequencies provide shorter skip distances. The higher frequencies are less readily refracted, so they do not bend back to Earth as quickly and return to Earth at a point further away from where the low-frequency signals return. Skip distance is therefore increased.

The relation of the frequency being transmitted and the refraction taking place in the ionosphere also shows up in the separation of the HF and VHF. Frequencies up to 30 MHz will refract and return to Earth. However, as the frequency increases, the refraction becomes less and less pronounced until finally there is no refraction and the signal penetrates the ionized band; this is called the *critical frequency*. This lack of refraction is a varying combination of both frequency and the density of the ionized later. The denser the ionization the more pronounced the refraction, and vice versa. The higher the frequency,

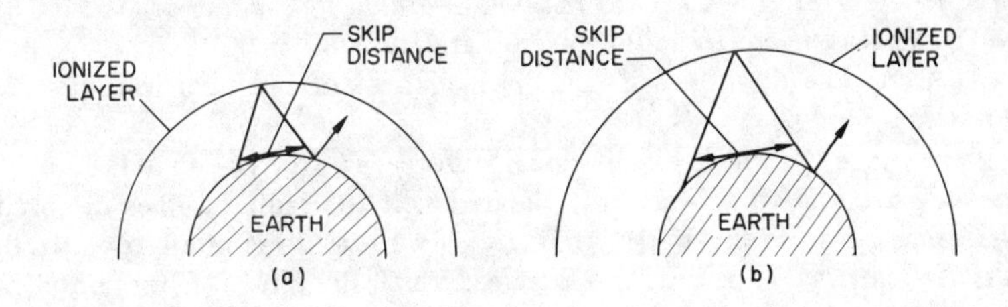

Fig. 23. Skip distance changes as a result of sunspot activity. (a) Skip distance with the ionized layer at normal height. (b) Increased skip distance due to ionized layer at an increased height as a result of sunspot activity.

the less the chance that there will be refraction. The dividing line is generally set at 30 MHz. This is a rule-of-thumb and the frequency can vary either way.

B.3.2(b) What are some of the propagation factors that influence radio transmission on the VHF amateur frequency bands?

Answer: VHF amateur frequency band radio transmission is affected by the lack of signal refraction, the availability of sporadic-E, and weather conditions.

Discussion: As noted in the discussion above, one of the dividing lines between HF and VHF is the point at which signals are no longer refracted but penetrate the ionized layers. The VHF bands are considered *local* frequencies, useful for *line-of-sight* operation between antennas with no useful skip distances involved.

On occasion, heavily ionized layers will appear at the same height, about 70 miles, as those of the E-band. Because of its sporadic appearance, it is referred to as *sporadic-E* ionization. Sporadic-E may appear at any time of day or night and when it does it will affect all bands, although it has its most pronounced effect on the 50-MHz band (6 m). Sporadic-E provides skip operation from 1000–2000 miles, which are skip distances not normally associated with 50 MHz.

Another major factor operating in the 50-MHz band is weather. During warmer periods or in the tropics, the contrasting cool-warm air masses create what are often referred to as ducts in which 50-MHz signals propagate over longer than normal distances.

The 144-MHz (2-m) band, as well as those bands above 220 MHz, is more consistent, since it is line-of-sight operation. Sporadic-E skip and ducting is rare.

B.3.3 What is meant by *maximum usable frequency*?

Answer: The highest frequency that can be used to operate over a specific distance before the critical frequency is reached.

Discussion: As noted in B.3.2(a), the critical frequency is that frequency just above the frequency that will reflect or refract from the ionosphere. Frequencies above the critical frequency do not return to Earth. If an operator desires to work a station 3000 miles away and finds that signals transmitted in the 10-m band are not received, whether due to being above the critical frequency or because they skip over the station being called, the frequency must be lowered to those of the 15-m band, or if necessary the 20-m band, etc. When reception is achieved at the desired distance using the highest possible frequency, this frequency is referred to as the *maximum usable frequency*, or MUF.

B.3.4(a) What is meant by radio wave polarization?

Answer: The direction of the lines of force of the electrical wave determine polarization.

Discussion: If the radiated lines of force are vertical in relation to the Earth, they are said to be vertically polarized. If the radiated lines of force are on the same plane as the Earth, they are said to be horizontally polarized. To radiate vertical waves the antenna is placed vertically, and for horizontal radiation the antenna is placed horizontally.

B.3.4(b) What is meant by radio wave spreading?

Answer: As the radio wave spreads out from the antenna, the wave intensity is inversely proportional to the distance from the antenna.

Discussion: The relationship between the distance from the antenna and the signal strength is said to be inversely proportional because the signal strength decreases as the distance increases. Thus, if at a specific distance from the antenna the signal strength is reduced to one half, then at twice the same distance the signal strength will be reduced to one quarter.

B.3.5 What is ionospheric absorption?

Answer: As radio waves travel through the ionosphere, they give up some of their energy when colliding with the ionized particles of the ionosphere; this is called absorption. Absorption is greatest at the lower frequencies, and varies directly with the intensity of the ionization.

C. Operating Procedures

C.3.1 What operating procedures can minimize interference and congestion of the amateur frequency bands?

Answer: One should listen carefully before transmitting so as not to interfere with other transmissions already in progress.

Only the power necessary for good signal reception, no more, should be used. When operating CW or phone, excessively long calls should not be made and proper calling, answering, and identification procedures should be used.

If available, a directive antenna should be utilized. The quality of the station emissions should be checked.

When tuning or testing a transmitter, the output should be applied to a dummy antenna. Use VHF or UHF frequencies for local communications.

Discussion: The operating procedures listed above are mostly self-explanatory. Antennas having directive characteristics provide a dual benefit. It concentrates the power in the direction of transmission, providing maximum reception in the same direction and reduces power radiation to the rear and sides of the antenna, which minimizes the possibility of interference in those directions.

The dummy antenna is a resistive load to which the output of the transmitter is connected. By applying the output power to the load instead of to the antenna, the signal is kept from radiating and possibly interfering with other signals. Depending upon the output power of the transmitter, the dummy load may be as simple as a light bulb or it may consist of a high-wattage resistor immersed in oil.

The use of 6 m and/or 2 m for local communications reduces the congestion in the lower frequency bands.

C.3.2 In what manner should an amateur radio station be operated in all respects not specifically covered by the Rules?

Answer: An amateur radio station should be operated in accordance with good engineering and good amateur practice. (See Section 97.77.)

C.3.3 How are harmonic frequencies avoided?

Answer: By tuning the rf amplifier stages using the minimum necessary drive; use high-Q tuned circuits tuned to the fundamental frequency; use a tuned antenna coupler (transmatch) between the transmitter output and the transmission line; ensure that all rf amplifier stages are properly shielded.

Discussion: Excessive drive signals cause signal clipping which results in harmonic frequency generation. A high-Q tuned circuit is selective, having a narrow bandwidth. A high-Q tuned circuit tuned to resonance will limit the bandwidth to only the desired frequencies. An antenna coupler provides additional tuned circuits that will block harmonic frequencies from passing to the antenna. Shielding prevents leakage of rf signals that might create undesired harmonic frequencies. Refer also to questions D.3.12 and F.3.2 below.

C.3.4 When may the Federal Communications Commission designate and specify what frequencies shall be used by amateur radio operators?

Answer: When the Commission, in its discretion, declares that a general state of communications emergency exists. (See Section 97.107, *Operation in emergencies,* for a full discussion of emergency operating procedures.)

D. Emission Characteristics

D.3.1 What system is used in the FCC rules for classifying and designating emissions from amateur radio stations?

Answer: Emission types are indicated by a system in which a letter designating the type of modulation used is followed by a number designating the method of transmitting the intelligence. For example, type A3 is amplitude-modulated (A), telephony (3).

Discussion: Subpart C, *Technical Standards,* Section 97.65, *Emission limitations,* discusses the types of emissions but makes no special note of the system used to designate emissions. The letters used to indicate the *type of modulation* are:

A – Amplitude
F – Frequency or phase
P – Pulse

The type of intelligence transmission is designated by:

0 – Absence of modulation used to carry intelligence
1 – Telegraphy, unmodulated
2 – Telegraphy, keyed on and off by an audio frequency
3 – Telephony
4 – Facsimile
5 – Television
6 – Four-frequency diplex telegraphy
7 – Multichannel voice-frequency telegraphy

Typical emission types used in amateur service include the following:

A0 – Steady unmodulated carrier
A1 – Telegraphy using continuous waves (CW)
A2 – Amplitude tone-modulated telegraphy
A3 – AM telephony including SSB
A4 – Facsimile
A5 – Television
F0 – Steady unmodulated carrier
F1 – Carrier-shift telegraphy
F2 – Audio frequency-shift telegraphy

F3 – Frequency (or phase) modulated telephony
F4 – FM facsimile
F5 – FM television

D.3.2 What are the characteristics and standards of good quality telephony emissions from amateur radio stations?

Answer: Good quality telephony emissions do not exceed 100% modulation (are not overmodulated), transmit only the speech portion of the audio spectrum, and, in single sideband, use linear amplification of the SSB signal.

Discussion: In amplitude modulation, the audio (speech) intelligence modulates (varies) the amplitude (height) of the station carrier signal. The carrier signal at radio frequencies is a high frequency signal compared to that of the modulating audio signal. When mixed with the carrier signal, the modulating audio signal, when positive, will *add* to the carrier signal, that is, it will add to *both* the negative *and* positive portions of the carrier signal. When negative, the modulating audio signal will *subtract* from *both* the negative and positive portions of the carrier signal.

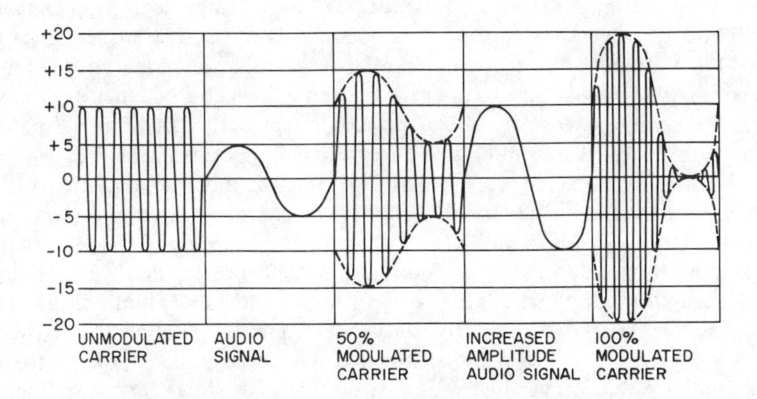

Fig. 24. Percentage modulation of a carrier.

Figure 24 illustrates mixing the unmodulated carrier with a modulating audio signal. When the modulating audio signal is one-half the amplitude of the carrier, the resulting modulated carrier is one-half again the amplitude of the unmodulated carrier. This 50% increase in amplitude is referred to as 50% modulation. When the modulating audio signal is equal in amplitude to that of the carrier, the positive and negative carrier swings are a 100% increase in those of the unmodulated carrier for 100% modulation.

When the modulating audio signal exceeds the amplitude of the carrier signal, the positive portion of the modulating signal will cause the positive and negative peaks of the carrier signal to rise to more than 100%. However, the negative portion of the modulating signal will cause the carrier to reach zero before the modulating signal has reached its most negative point. The resulting modulated carrier signal, shown in Fig. 25, is distorted. This is *overmodulation.*

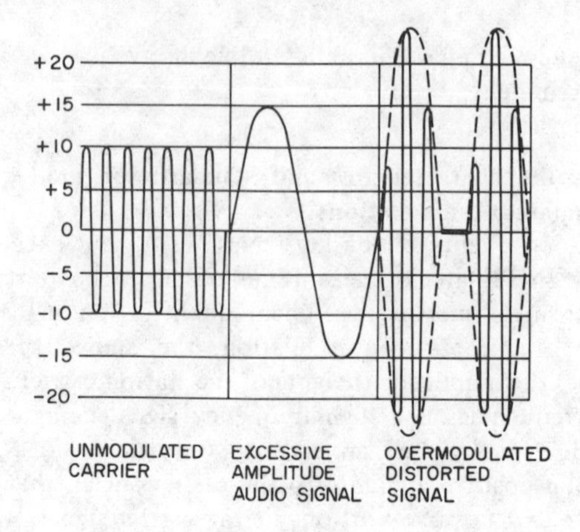

Fig. 25. Overmodulated signal showing resulting distortion.

The total audio spectrum is considered to range from approximately 30–20,000 Hz. Speech frequencies fall mainly in the 200–3000-Hz range with 300–3000 Hz considered to be the central or main range of speech frequencies. Special microphones whose response is limited to speech frequencies are used to obtain maximum output at these speech frequencies. This limited frequency response also permits simplified amplifier design and construction.

When a single sideband signal is developed, it must be amplified without adding any distortion, since the SSB signal is sensitive to amplifier distortion. To prevent distortion SSB, amplifiers must be linear amplifiers, meaning that the power output will be exactly proportional to the power input. This linearity permits amplifying only the SSB signal without adding any amplifier distortion.

The types of distortion that may develop in a nonlinear amplifier are amplitude, frequency and phase distortion. Frequency distortion exists where the amplifier either discriminates against or overly amplifies certain frequencies. Phase distortion results when undesired phase shift is introduced to place some of the frequencies out of phase with others. Amplitude distortion occurs when portions of the signal receive more, or less, amplitude gain than do other portions of the signal.

D.3.3 What range of audio frequencies is usually adequate for excellent voice intelligibility in communication systems?

Answer: The range of 300–3000 Hz. (See preceding discussion for D.3.2.)

D.3.4(a) How is voice information conveyed in amplitude-modulated emissions?

Answer: The amplitude of the speech signal directly varies the amplitude of the carrier signal, thus conveying voice information. (See D.2.1.)

D.3.4(b) How is voice information conveyed in frequency-modulated emissions?

Answer: The speech frequencies directly vary the carrier frequency above and below its center frequency. The stronger the amplitude of the speech signal the further the carrier frequency swings from its normal center frequency.

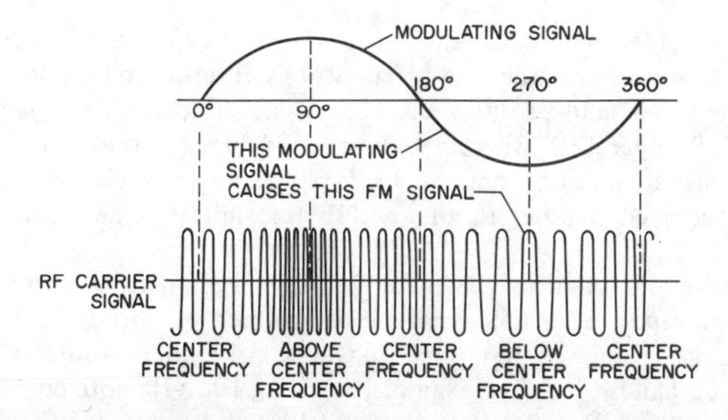

Fig. 26. Amplitude variations of the modulating signal cause frequency variations in the carrier wave.

Discussion: The effect of speech signal on the frequency of the carrier causes frequency-modulated emission. As shown in Fig. 26, as the speech modulating signal goes positive, the carrier frequency increases, reaching a maximum at the peak positive amplitude of the modulating signal. As the speech modulating signal goes negative, the carrier frequency decreases, reaching minimum at the peak negative amplitude of the modulating signal. The amount of carrier frequency deviation from the center frequency is proportional to the amplitude of the speech modulating signal.

D.3.4(c) How is voice information conveyed in phase modulated emissions?

Answer: Voice information is conveyed in phase modulated emissions by having the speech amplitude vary the phase of the carrier signal. This phase change in turn creates a frequency change in the carrier similar to that of frequency-modulation.

D.3.5 What is meant by the occupied bandwidth of an emission?

Answer: The occupied bandwidth is the total bandwidth of frequencies occupied by the carrier and sidebands as determined by the type of emission.

Discussion: The method of modulating the carrier determines the frequencies developed on either side of the carrier. These *sidebands* take up frequencies on one or both sides of the carrier, occupying a frequency band around the carrier. The total bandwidth developed is the "occupied bandwidth" of an emission.

A CW signal essentially consists of only an interrupted carrier, implying that there is no bandwidth. There is, however, a slight bandwidth averaging about 100 Hz, which is related to the speed of sending. Thus, a typical CW emission has an occupied bandwidth of about 100 Hz.

An amplitude-modulated (AM) signal provides sidebands on both sides of the carrier, which equal the value of the modulating frequency. For example, a 1000 Hz note provides sideband frequencies 1000 Hz above and below the carrier. Assuming that the speech frequencies are limited to 3000 Hz as noted in D.3.3 above, the maximum sideband frequencies would be 3000 Hz on either side for a total bandwidth of 6000 Hz. Thus an AM signal may have a maximum occupied bandwidth of 6 kHz.

The two sidebands of an AM signal are a mirror image of each other. To detect an AM signal, one sideband is removed by filtering and the intelligence of the remaining sideband is utilized. This waste of a sideband led to single sideband (SSB) operation wherein only one sideband is transmitted. A 3-kHz modulating signal develops only a 3-kHz sideband from an SSB transmitter. Thus, the occupied bandwidth of an SSB transmitter is half that of an AM transmitter.

Frequency-modulation (FM) bandwidth determination is complex. The sidebands developed in an FM signal are determined by both the amplitude and the frequency of the audio-modulating signal. Frequency-modulated signals require a wide bandwidth. For example, a strong 1000-Hz note could develop 8 sidebands on either side of the carrier for a 16-kHz bandwidth. Because of this wide occupied bandwidth of an FM signal, the use of wideband FM has been restricted to the VHF bands. The bandwidth available in the VHF amateur frequency bands is wider than those of the HF bands.

D.3.6(a) What is wideband F3 emission?

Answer: F3 wideband emission refers to frequency-modulated radiotelephony. Sidebands for wideband F3 may be no wider than 15 kHz on either side of the carrier for an occupied bandwidth of 30 kHz.

D.3.6(b) What is narrowband F3 emission?

Answer: F3 narrowband emission refers to frequency-modulated radiotelephony in which the sidebands may be no greater than those of an AM emission, 3 kHz either side of the carrier, for an occupied bandwidth of 6 kHz.

D.3.6(c) On what amateur frequencies may each (wideband F3 and narrowband F3) be used?

Answer: Wideband F3: 52.5–54.0 MHz; 144–148 MHz and all amateur bands above these frequencies.

Narrowband F3: 3.775–4.0 MHz; 7.075–7.10 MHz; 7.15–7.30 MHz; 14.20–14.35 MHz; 21.25–21.45 MHz; 28.50–29.70 MHz; 50.1–52.5 MHz.

D.3.7 What is the maximum percentage of modulation permitted by the rules? What does the term mean?

Answer: The maximum percentage permitted is 100%. Maximum percentage of modulation means that the amplitude of the modulating audio signal equals the amplitude of the carrier signal for full, or 100%, modulation. (See the discussion for D.3.2.)

D.3.8(a) What is peak-envelope power (PEP) in an emission from an rf amplifier?

Answer: At the maximum or peak value of the carrier emission, the peak carrier voltage and current provide the peak power. The rms (root-mean-square) value (0.707 of peak value) of this power is the peak envelope power (PEP).

Discussion: As shown in Fig. 27, the peak of the carrier waveform (envelope) is the highest when the carrier amplitude is at maximum. At this peak value, the level of carrier voltage and current is highest. The term peak can be misleading because, although peak values are measured, the PEP is determined by using the rms value of these peak readings.

The rms value of an alternating current is determined by a complex method. The instantaneous value of current during one cycle is constantly changing from zero to maximum throughout the cycle. A cycle is a circle and,

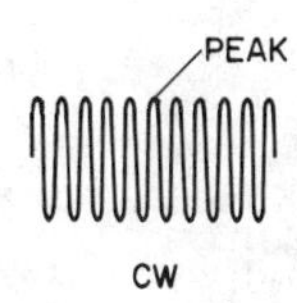

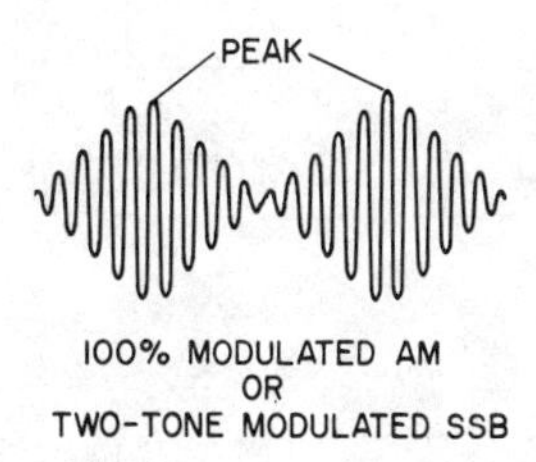

Fig. 27. Carrier amplitude peaks remain the same for CW and peak at the highest value of modulating signal for AM and SSB.

therefore, can be divided into 360 parts, each of which is called a degree. Each 1/360th of a cycle (each degree) has an instantaneous value. Each value is squared and then the average value of all the squared values is found. The last step is to find the square root of the average value. This is known as finding the *root-mean-square* value. For pure sine waves the rms value will be 0.707 of the maximum or peak value.

The peak power and average power are identical for a CW signal, since the power remains constant during transmission. Peak power for a CW emission is the rms value of carrier current times the rms value of carrier voltage.

For a 100% modulated AM signal, the power of the modulating signal adds to that of the carrier signal. Thus, at the peak of the envelope, the carrier voltage is added to the modulator voltage and the carrier current is added to the modulator current. Since both the current and voltage are doubled, the power at the peak of a 100% modulated AM signal is four times that of an equivalent unmodulated carrier.

Modulating an SSB transmitter with two equivalent amplitude audio frequencies (two tones), each slightly different in frequency (typically 1000Hz apart), provides an SSB output signal similar in appearance, when viewed on an oscilloscope, to that of the 100% modulated AM signal. Using a two-tone audio input is the standard method of applying a test output signal to check an SSB transmitter. This provides a familiar waveform to check. PEP output will be double the rms voltage and current value, PEP = $2\ (E_{rms} \times I_{rms})$, because there are two modulating signals.

Review: The PEP output for CW is the value of the rms voltage times the value of the rms current; for AM, the PEP output is four times the value of the unmodulated carrier signal, and for SSB it is twice the value of the unmodulated carrier signal.

D.3.8(b) How can the PEP emission from an rf amplifier be determined?

Answer: PEP emission can be determined by using a peak-reading rf wattmeter, an rf voltmeter, or an oscilloscope.

Discussion: Peak-reading rf wattmeters are relatively costly commercial devices not normally available to amateurs. A peak-reading wattmeter is placed in the transmission line between the output of the SSB transmitter and the antenna (or dummy load) to monitor the peak value of the emission. If the readings are made for testing purposes, using a two-tone generator, the output should be applied to a dummy load to prevent interference with other transmissions.

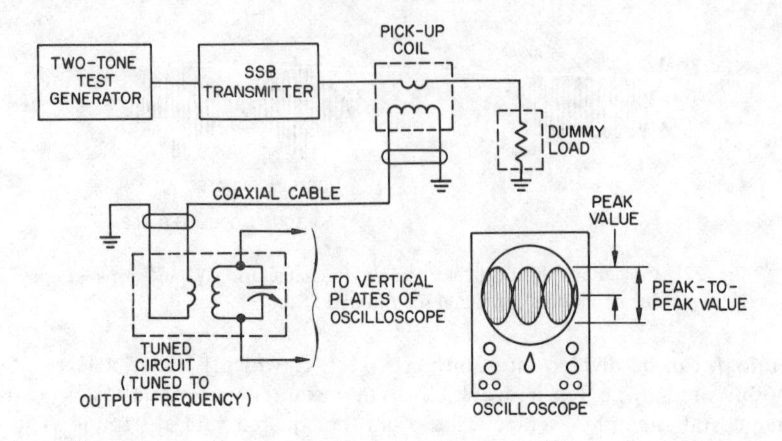

Fig. 28. Using a calibrated oscilloscope to obtain an output pattern of an SSB transmitter to check the amplitude of the output signal.

By applying the modulated output of the SSB transmitter to a resistive dummy load the rms value of the rf voltage across the load is measured with an rf voltmeter. The formula, PEP = $E_{rms}{}^2/R$, is then used to calculate the PEP emission.

A method more readily available to amateurs is the use of a pickup coil coupled to a coil in series with the transmission line (Fig. 28). Because this test requires a two-tone test generator, the output of the transmitter should be applied to a dummy load to prevent interference with other transmissions. The output of the pickup coil is applied to a circuit tuned to the transmitter frequency. The signal voltage developed in the tuned circuit is then applied to the vertical plates of the cathode ray tube (CRT) in an oscilloscope. Setting the internal sweep circuit of the oscilloscope for approximately 250 Hz will provide a pattern on the face of the CRT similar to that shown in Fig. 28. Note that it is the same as a 100% modulated AM signal.

The oscilloscope is calibrated to measure the output waveform. To calibrate an oscilloscope, one determines how high the amplitude of a known value voltage is when displayed on the face of the CRT. By comparing the peak amplitude of the transmitted output signal with that of the calibration voltage, the amplitude of the emitted signal can be determined. Knowing the peak output voltage and the impedance of the dummy load, the PEP can be calculated by using the rms value of the peak voltage in the formula, PEP = $E_{rms}{}^2/Z_{load}$.

D.3.8(c) How can the PEP input to the final amplifier stage(s) supplying power to the antenna be determined?

Answer: The PEP input is determined by multiplying the metered value of the plate voltage by the highest metered value of the plate current of the final amplifier stage(s) while a modulated signal is being emitted.

Discussion: PEP input refers to the peak values. A meter pointer cannot follow the voice fluctuations; therefore, it will tend to give an average value reading which is considered proper for PEP input. Inexpensive meters may respond in a sluggish manner to the modulation-caused fluctuations and can give an erroneous low average value. To prevent this, the FCC has specified that the meter time-constant shall not exceed 0.25 sec. This means that the meter reading must reach no less than 63% of the total applied voltage (or current) in 0.25 sec or less.

D.3.9(a) What is average power in an emission from an rf amplifier?

Answer: Average power is the average value of all the amplitude values of each cycle taken over a sufficient period of time.

Discussion: To find the average value of ac voltage, multiply the peak value by 0.636. (Note the relative closeness to the rms value of 0.707.)

D.3.9(b) How can the average power in an emission from an rf amplifier be determined?

Answer: The average power can be determined by using of an rf voltmeter or an oscilloscope.

Discussion: Refer to the discussion of D.3.8(b) above. By measuring the peak voltage value and multiplying it by 0.636, the formula, $P_{avg} = E_{avg}^2 / Z_{load}$, can be used.

D.3.9(c) How can the average power input to the final amplifying stage(s) supplying power to the antenna be determined?

Answer: The average power input can be determined by multiplying the metered values of plate voltage by the highest reading value of plate current of the final amplifier stage(s) while a modulated signal is being emitted.

Discussion: As noted in the discussion of D.3.8(c), the meter pointer can provide only an average value reading of the fluctuations of a modulated signal.

D.3.10 In a single sideband, suppressed carrier A3 emission (SSB) transmitter, what determines the PEP-to-average power ratio? What are typical values?

Answer: The waveform produced by the modulating signal determines the PEP-to-average power ratio. Typical values are 2:1 and 3:1.

Discussion: As shown in Fig. 29, the difference between the peak values (which determines PEP) and the average value of an SSB signal will vary with both the syllables being pronounced and the characteristics of the voice. The figure illustrates the typical ratios of 2:1 and 3:1.

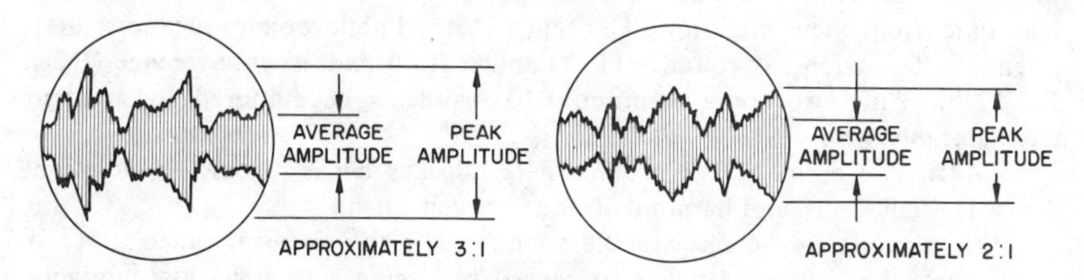

Fig. 29. Typical values of PEP-to-average ratios determined by the modulating waveforms.

The average voice modulation pattern is considered to have a 2:1 PEP-to-average power ratio, which permits a maximum 2000-W PEP input on peaks of a voice modulating signal on SSB emission, since the 2:1 ratio will average this to the same maximum value as the 1000-W input permitted for an unmodulated (CW) carrier.

D.3.11(a) What does the term *S/D ratio* in a SSB transmitter mean?

Answer: The S/D ratio is the signal-to-distortion ratio of an SSB transmitter.

Discussion: The S/D ratio, measured in decibels (dB), is the ratio of the peak carrier voltage to the value of voltage developed by a distorted signal. This intermodulation distortion is a result of undesired nonlinearity in a linear amplifier, such as that used in an SSB transmitter. Application of different frequency signals to a linear amplifier having some degree of nonlinearity creates sum and difference frequencies that are called intermodulation distortion products. These products are identified by successive peak values. The third peak value, for example, is called the third-order product. For the S/D ratio of a SSB transmitter, the third- and fifth-order intermodulation products are measured and compared to the peak carrier product. A high value ratio indicates very low levels of intermodulation distortion in the SSB transmitter linear amplifier. A high ratio of S/D is desirable, since the higher the S/D ratio the greater the available PEP.

D.3.11(b) How can the S/D ratio be determined?

Answer: The S/D ratio is determined by the use of test instruments such as a spectrum analyzer or a distortion analyzer which determines the intermodulation distortion products of an SSB transmitter with a two-tone test signal applied. These products can then be compared with the peak carrier voltage as determined by a peak reading rf voltmeter, or an oscilloscope, to determine the S/D ratio. Another method is the use of a highly selective SSB receiver capable of tuning to the carrier, also to the two detected tone frequencies applied to the transmitter, and to the third- and fifth-order intermodulation products that can be detected by using the receiver S-meter for calibrated dB readings.

D.3.12(a) What is RFI?

Answer: RFI is radio frequency interference, which is caused by undesired transmitter emissions.

Discussion: Undesired rf radiation (also called spurious emissions) is capable of creating interference both with amateur communications and with public communications. Amateur communications interference can result from harmonic frequency radiation, key clicks, etc. Public communication interference is television interference (TVI) and radio broadcast interference (BCI).

D.3.12(b) What are some common RFI problems encountered by amateur radio stations?

Answer: Common RFI problems include key clicks, backwave radiation, parasitic oscillations, and harmonic frequency generation.

Discussion: Key clicks are the sounds heard in a receiver tuned to a CW signal containing spurious radiation caused by arcing across the key contacts. Backwave radiation is caused by the leaking of some of the oscillator signal of a CW transmitter, which is transmitted while the key is in the up or open position. Parasitic oscillations occur when an inductance resonates with stray capacity to break into oscillation and create interfering signals that are unknowingly transmitted. The inductance can be that of a length of wire; it need not be an inductor itself such as an rf choke.

Harmonic frequencies are generated in tuned circuits, particularly the Class C tuned circuits used in transmitters. These undesired harmonics develop signals sufficient in strength to interfere with other amateurs and with other assigned frequency bands such as television channels. Harmonic frequencies are multiples of the original frequency. As the multiple increases, the harmonic becomes weaker, that is, the second harmonic is weaker than the fundamental frequency, the third harmonic is weaker than the second harmonic, and so on.

D.3.12(c) What are the solutions to the RFI problems?

Answer: To remove key clicks, a filter is used across the key contacts. To remove backwave radiation, the oscillator buffer and driver stages should be shielded.

To prevent or eliminate parasitic oscillations, all signal leads should be shielded or located so as to prevent feedback; low value noninductive resistors should be placed in series with grid leads, and plate leads should have a low value noninductive resistor with three or four turns of wire wrapped around the resistor and placed in series with the plate lead.

To prevent excessive harmonics, the circuit must be properly tuned and the tuned circuit must have a high enough value of Q (quality factor) to prevent the generation of harmonics. An antenna coupler is used between the transmitter final amplifier output and the transmission line; the rf amplifier stages are shielded, and the lowest value of grid drive is used to the tuned amplifier circuit necessary to obtain the desired output.

Discussion: A basic key click filter used to remove the arcing from opening key contacts appears in Fig. 30.

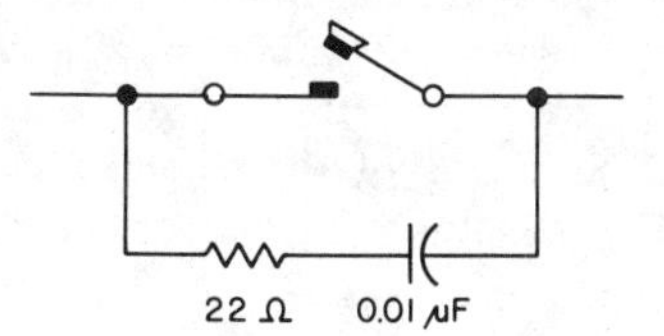

Fig. 30. A key-click filter.

For prevention of harmonic generation, these precautions are necessary: Excessive grid bias and/or excessive signal drive to a typical Class C rf amplifier causes the input signal to be clipped and appear similar to a square wave pulse. Square waves are rich in odd-order harmonics. Although Class C requires a high value of bias and signal drive, only the minimum value needed should be used.

The Q of the tuned rf amplifier resonant circuit is important in determining the generation of excessive harmonics. The Q of a resonant inductive-capacitive circuit is found by dividing the reactance of the circuit by the resistance of the circuit, $Q = X/R$. At resonance, the inductive and capacitive impedances are canceled and the determining factor for Q is the dc resistance offered by the circuit. A circuit with a low value resistance compared to the values of reactance is a high-Q circuit. For example, if at resonance the reactance offered by the inductance (or capacitance) were 250 Ω and the resistance of the circuit were 25 Ω, the Q would be 10, for a low-Q. If in the same circuit the resistance at resonance were to be 2.5 Ω, the Q would be 100, for a high-Q. A low-Q circuit has low gain and a wide bandwidth; a high-Q circuit has higher gain and a narrower bandwidth (Fig. 31).

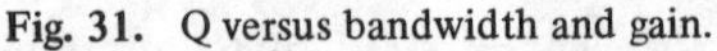
Fig. 31. Q versus bandwidth and gain.

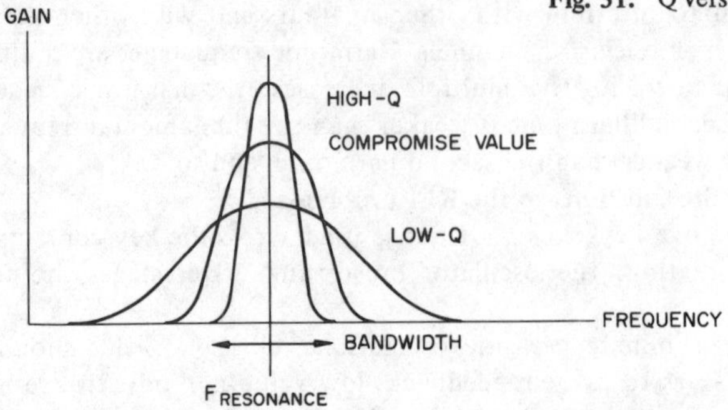

The major method of harmonic reduction is the frequency filtering effect of the tuned circuit. A low-Q resonant circuit has a wider bandwidth than that of a high-Q circuit. A high-Q resonant circuit can be so narrow in bandwidth that it cuts off some of the desired frequencies, causing high losses. A compromise is usually sought in which the Q of the resonant circuit provides a bandwidth that is high enough to prevent undesired harmonics, yet not so high as to restrict the bandwidth and provide high losses.

An antenna coupler placed between the output of the transmitter and the transmission line accomplishes two things: It provides additional tuned circuits to prevent the passage of harmonics; and it helps to match the transmitter output to the transmission line to keep the transmission line losses to a minimum, thus providing maximum power output.

Shielding the rf amplifier stages is a method of containing harmonics within the physical area of the shielded stage. This prevents harmonics from interfering with other stages of the transmitter.

E. Electrical Principles

E.3.1(a) How do resistors combine in parallel and series?

Answer: The combined total resistance of two or more resistors in parallel will be less than that of the lowest value resistor. The combined total resistance of two or more resistors in series will be the sum of the combined individual resistor values.

Discussion: Resistors in parallel (Fig. 32a) divide the flow of current by offering two or more alternate paths. This division of current allows more

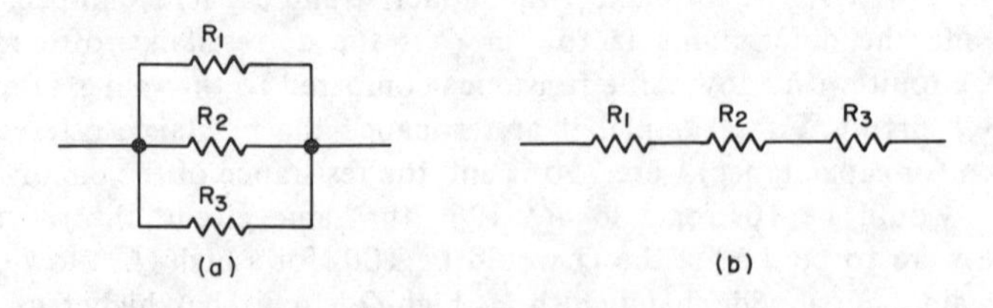

Fig. 32. Resistors in parallel (a) and in series (b).

current to flow through the circuit. To find the total resistance of two resistors in parallel, the following formula is used:

$$R_{\text{total}} = \frac{R_1 \times R_2}{R_1 + R_2}$$

For two or more resistors in parallel, the following formula is used:

$$R_{\text{total}} = \frac{1}{\frac{1}{R_1} + \frac{1}{R_2} + \frac{1}{R_3} \text{ (etc.)}}$$

Resistors placed in series (Fig. 32b) cause the current to flow through each resistor, one after the other. Each resistor offers opposition to the flow of current. The increased total resistance reduces the flow of current. The total resistance of two or more resistors in series is the sum of the values of each resistor. This is expressed as $R_{\text{Total}} = R_1 + R_2 + R_3$ (etc.).

E.3.1(b) How do capacitors combine in parallel and series?

Answer: The combined total capacitance of two or more capacitors connected in parallel is the sum of the combined individual capacitors. The combined total capacitance of two or more capacitors in series is less than that of the lowest value capacitor.

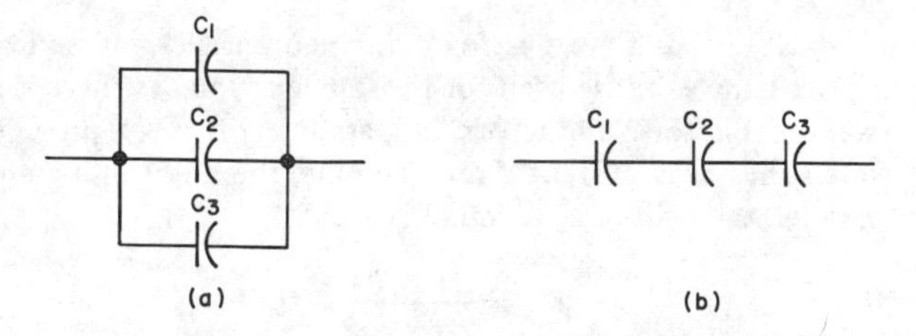

Fig. 33. Capacitors in parallel (a) and in series (b).

Discussion: Capacitors in parallel (Fig. 33a) in effect increase the plate area by combining the total plate area of the two or more capacitors. To find the total capacitance of two or more capacitors in parallel, values of each capacitor are added, the sum value being the total capacitance. This is expressed as $C_{\text{Total}} = C_1 + C_2 + C_3$ (etc.).

With capacitors connected in series, the dielectric insulating material of each capacitor is placed in series. The capacitance value is reduced when the plates are further separated by a thicker dielectric; placing capacitors in series reduces the value of capacitance to less than that of the lowest value capacitor. To find the total value of capacitance for two capacitors in series, the following formula is used:

$$C_{\text{total}} = \frac{C_1 \times C_2}{C_1 + C_2}$$

For two or more capacitors in parallel, the following formula is used:

$$C_{\text{total}} = \frac{1}{\frac{1}{C_1} + \frac{1}{C_2} + \frac{1}{C_3} \text{ (etc.)}}$$

E.3.1(c) How do inductors combine in parallel and series?

Answer: The combined total inductance of two or more inductors in parallel is less than that of the lowest value inductor provided there is no mutual coupling between inductors. The combined total inductance of two or more inductors in series is the sum of the combined individual inductors provided there is no mutual coupling between inductors.

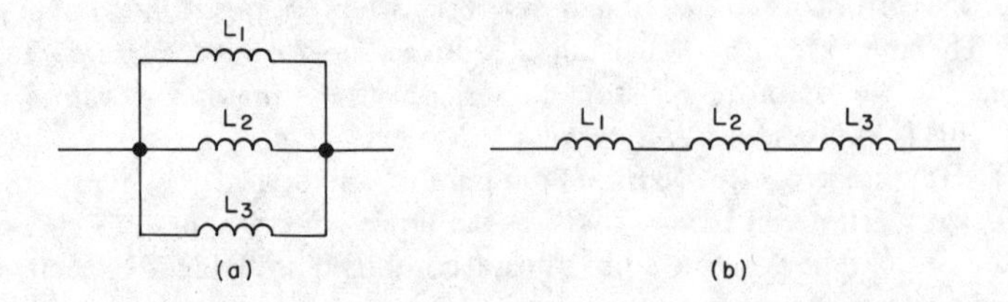

Fig. 34. Inductors in parallel (a) and in series (b).

Discussion: Note that the rules for combined parallel and series values of inductance are the same as those used for resistance as long as there is no mutual coupling between inductors. Inductors in parallel (Fig. 34) provide reduced inductance which increases current flow. To find the total inductance of two inductors in parallel, the following formula is used:

$$L_{\text{total}} = \frac{L_1 \times L_2}{L_1 + L_2}$$

For two or more inductors in parallel, the following formula is used.

$$L_{\text{total}} = \frac{1}{\frac{1}{L_1} + \frac{1}{L_2} + \frac{1}{L_3} \text{ (etc.)}}$$

In calculating the total value of inductance of two or more inductors, it is assumed the inductors are not so close to each other as to have their magnetic fields couple with each other. Should the magnetic fields of two or more inductors be mutually coupled, the values of inductance will be changed as determined by the degree and the direction of coupling.

E.3.2(a) How does voltage division occur across series-connected resistors?

Answer: The voltage divides across each resistor as determined by the value of each resistor and the value of current flow in the circuit. The total of the voltage drops across the individual resistors equals the value of applied voltage.

Discussion: The value of applied voltage will force a current to flow through the resistors connected in series. The value of current flow is determined

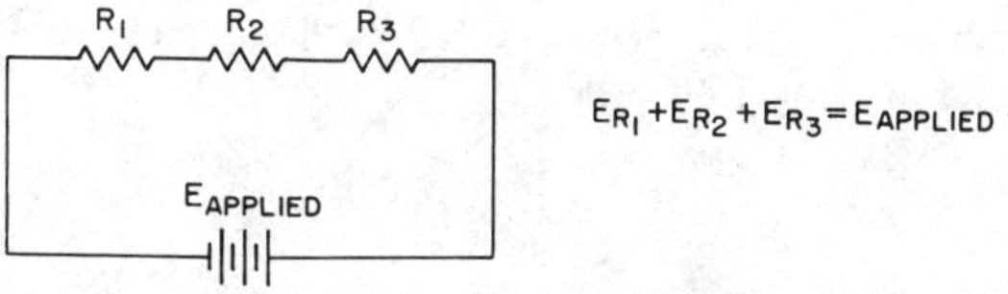

$E_{R_1} + E_{R_2} + E_{R_3} = E_{APPLIED}$

Fig. 35. Voltage division in series-connected resistors.

by the value of total resistance and the value of voltage being applied, ($I = E_{applied}/R_{total}$.

Since the current flow in a series circuit is the same throughout the circuit, the value of current flow in each resistor is the same. Knowing the value of current flow and the value of resistance, the voltage across each resistor can be found by using $E = I \times R$. The voltage drops across each series-connected resistor will total the value of the applied voltage (Fig. 35).

E.3.2(b) How does voltage division occur across series-connected capacitors?

Answer: The value of applied voltage divides across each capacitor as determined by the value of the capacitor. The voltage is inversely proportional to the capacitance. The total of the voltage drops across the capacitors will equal the value of applied voltage.

Discussion: There are two methods of calculating the voltage drops across series-connected capacitors. One method uses the reactance of a capacitor to ac voltage, as described in E.3.3(b) below. Capacitive reactance is the opposition to the current flow through a capacitor. The voltage divides across each capacitor as determined by the reactance. The capacitors with the least reactance have the lowest voltage drops. If the voltage and frequency are held constant, capacitive reactance will *decrease* with an *increase* in capacity. Therefore, the voltage drop across a larger value capacitor is less than that of a smaller value capacitor in a series circuit. In other words, the voltage across a series-connected capacitor is inversely proportional to its capacitance.

Inverse proportion is the opposite of direct proportion. For example, in a directly proportional relationship, such as resistance value to voltage drop, doubling the value of the resistance will double the voltage drop across a resistor. However, in an inversely proportional relationship, such as capacitor size to voltage drop, doubling the size of a capacitor will *halve* the voltage drop across the capacitor.

Another way to determine the voltage division across a series-connected capacitor is first to determine the total capacitance of the series-connected capacitors as described in E.3.1(b) above. The voltage across each capacitor can be found by using the following formula:

$$E_{capacitor} = \frac{C_{total}\ (\mu F)}{C_{individual}\ (\mu F)} \times \text{applied voltage}$$

For example, in a circuit with two capacitors in series, where one is 2 μF and the other is 4 μF and the total applied voltage is 400 V, the individual capacitor voltage is determined by first calculating the total capacitance of the circuit.

The value of the total capacitance is then substituted in the formula above to determine the voltage across each capacitor:

$$C_{\text{total}} = \frac{C_1 \times C_2}{C_1 + C_2} = \frac{2 \times 4}{2+4} = \frac{8}{6} = 1.33\ \mu\text{F}$$

$$E_{\text{capacitor 1}} = \frac{1.333}{2} \times 400 = 266.4\ \text{V}$$

$$E_{\text{capacitor 2}} = \frac{1.333}{4} \times 400 = 133.2\ \text{V}$$

The total voltage drop across the capacitors is equal to the applied voltage. (The 0.4-V difference in this case is due to rounding off the decimal value.) Note that the smaller 2-μF capacitor has twice the voltage of the larger value 4-μF capacitor because the capacitor size is inversely proportional to the voltage drop.

E.3.2(c) How does voltage division occur across series-connected inductors?

Answer: The voltage divides across each inductor in proportion to the value of each inductor and the value of current flow in the circuit. Again, the total of the voltage drops across the inductors equals the value of the applied voltage. This assumes no mutual coupling between inductors.

Discussion: The rules for determining the voltage division across series-connected inductors are the same as those for resistances. This is true if each inductor is separated from the other inductors so that there is no mutual coupling between inductors. (Refer to the discussion of E.3.1(c) above.)

Inductors, as capacitors, provide a reactance to ac voltage. This inductive reactance (X_L) is described in E.3.3(a) below. By using the value of X_L in place of the value of R, the same formulas apply to determining the values of series-connected inductors as for series-connected resistors explained above in E.3.2(a).

E.3.3(a) What is inductive reactance?

Answer: Inductive reactance (X_L) is the measurement (in ohms) of the opposition of an inductor to an alternating current.

Discussion: Inductance is the property of a wire carrying alternating current in which the expanding and contracting magnetic fields induce a voltage (back EMF) that opposes the changing values of current flow. Opposition to alternating current is called reactance (X) and is measured in ohms.

A series of turns or loops of wire form an inductor (L). The magnetic fields around these loops reinforce each other so that a coil offers a more substantial value of inductance than a straight wire. The reactance offered by an inductor is referred to as inductive reactance and is symbolized by X_L.

A coil has both reactance and resistance. The resistance is offered by the turns of wire of the inductor to direct current and the reactance is offered by the inductor to alternating current. As a rule, the dc resistance is so small in comparison to reactance that it is ignored in computing reactive values of a circuit. The value of inductive reactance is found with the formula $X_L = 2\pi fL$, where f is the frequency in hertz and L is the inductance in henries. Inductive reactance is directly proportional to frequency, that is, as the frequency increases the reactance also increases.

Inductive reactance causes the current flow to be impeded. As a result, the current flow follows or lags, the voltage. In a pure inductive circuit, current lags the voltage by 90°.

E.3.3(b) What is capacitive reactance?

Answer: Capacitive reactance (X_C) is the measurement in ohms of the opposition of a capacitor to an alternating current.

Discussion: A capacitor consists of two conducting plates separated by an insulator. A charge develops across the capacitor when current is applied. As the charge of voltage across the capacitor increases, it opposes the voltage being applied to the capacitor. The opposition offered by the capacitor is the reactance of a capacitor and is labeled X_C.

The value of capacitive reactance is found by using the formula, $X_C = 1/2\pi fC$, where f is the frequency in hertz and C is the capacitance in farads. The fact that the value of $2\pi fC$ is divided into 1 indicates it is an inverse proportion. Since most capacitor values are in microfarads, (millionths of a farad) a simpler formula is

$$X_C = \frac{1,000,000}{2\pi fC(\mu F)}$$

Capacitive reactance causes the voltage to be impeded; as a result, the voltage follows, or lags, the current. In a pure capacitive circuit, voltage lags the current by 90°.

E.3.3(c) What is impedance?

Answer: Impedance (Z) is the total opposition, measured in ohms, of a circuit containing both reactance and resistance.

Discussion: A circuit containing both reactance and resistance will affect the current flow in dc and ac circuits differently. The ohmic resistance of the wire forming the inductor is the only opposition an inductor offers to the flow of dc current. In an ac circuit, the inductor offers high opposition in the form of reactance. A capacitor is an open circuit to dc, and also offers reactance to ac.

In a circuit containing both X_L and X_C, reactances differ in their action; the combined action of X_L and X_C is referred to as circuit impedance. Frequently, the opposition offered to the flow of current in a circuit is called circuit impedance regardless of whether the opposition is resistive only, resistive and inductive reactance, resistive and capacitive reactance, or inductive and capacitive reactances.

In a resistive circuit, current and voltage flow together, that is, current and voltage are *in phase.* In circuits containing inductive reactance, the opposition offered to current flow causes the current to lag behind the voltage by 90°. In circuits containing capacitive reactance, the opposition to the applied voltage causes the voltage to lag by 90°. This is illustrated vectorially in Fig. 36. As an aid in remembering where current or voltage lead or lag, the student may wish to use the expression "*ELI* the *ICE* man." *ELI* indicates the voltage (E) in an inductor (L) is ahead of, or leads, the current (I). *ICE* indicates that the current (I) in a capacitor (C) leads the voltage (E).

The total reactance of a series circuit containing both inductive and capacitive voltages cannot be found by adding the resulting reactances, because

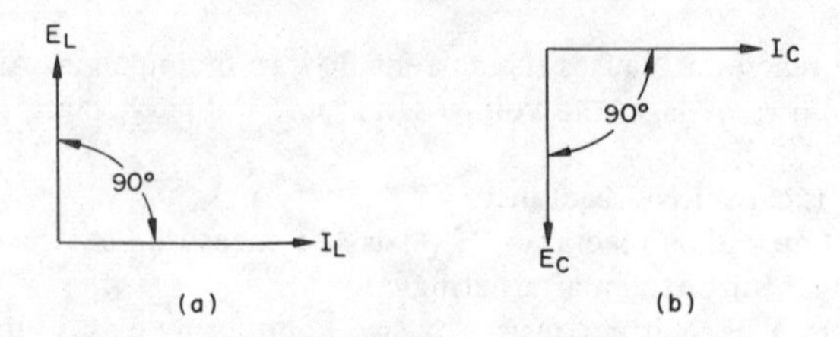

Fig. 36. Vectorial representation of the relation of current and voltage in circuits containing inductive and capacitive reactance. (a) In an inductor, E leads I by 90°. (b) In a capacitor, E lags I by 90°.

capacitive voltage and inductive voltage and their resulting reactances act in opposite directions as shown vectorially in Fig. 36.

To find the total opposition (impedance, Z) in a circuit containing both inductive and capacitive reactances, the vectors can be combined, since the current vector for both reactances is a common factor, as shown in Fig. 37(a). In the vector, E_L and E_C have been replaced with X_L and X_C, respectively. Since X_L and X_C are of opposite direction, they oppose and cancel each other.

In a circuit where X_L predominates, the resulting reactance of the circuit is inductive, and the circuit represents a value of X_L, shown in Fig. 37(b). The opposite action resulting in a circuit representing X_C is shown in Fig. 37(c).

To review. A circuit containing X_C and X_L offers a total reactance, which is determined by subtracting the reactive values. The final reactance value is that of the greater reactance. In a circuit containing resistance, voltage (E) and current (I) are in phase at all times. In a circuit containing X_L and X_C and R, the vector representation is drawn with X_C and X_L being opposite each other and

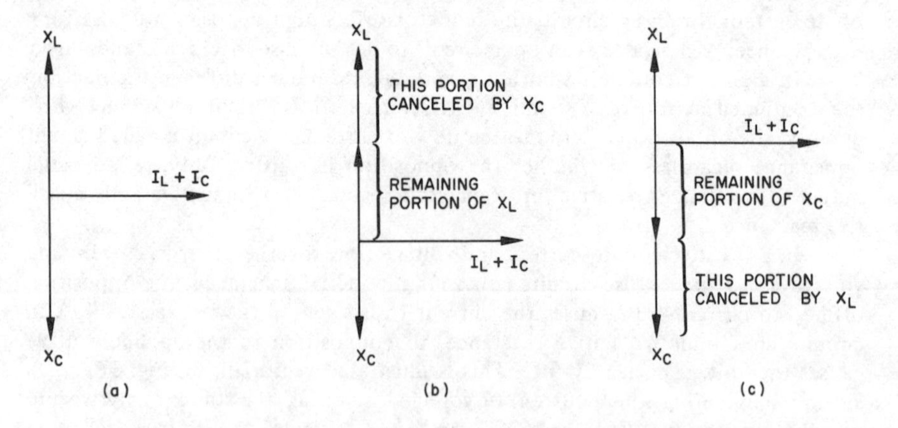

Fig. 37. Vectorial representation of resulting reactances in a series circuit containing X_C and X_L. (a) Combining the current and reactance vectors of inductive and capacitive reactances. (b) The resulting reactance of a circuit containing more X_L than X_C. (c) The resulting reactance of a circuit containing more X_C than X_L.

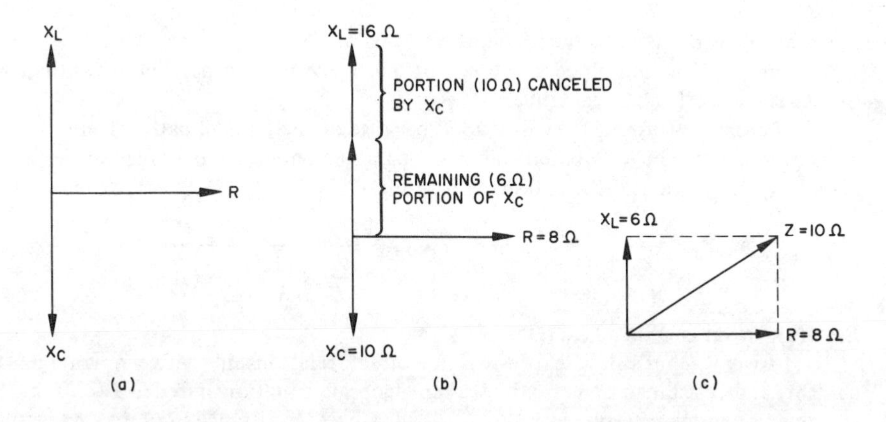

Fig. 38. Vectorial representation of impedance (Z) in a series circuit containing X_C, X_L, and R. (a) Vector diagram of X_L, X_C, and R in a series circuit. (b) Resulting reactance of a circuit containing more X_L than X_C. (c) Vector diagram and parallelogram showing resulting impedance.

with R replacing I (Fig. 38a). R replaces I in the vector diagram, because in a resistance, E and I are in phase.

As an example, the student can consider a series circuit containing X_L of 16 Ω, X_C of 10 Ω, and R of 8 Ω, as shown vectorially in Fig. 38(b). Since $X_L - X_C$ provides a remaining value of X_L of 6 Ω, the circuit can be redrawn as shown in Fig. 38(c). Figure 38(c) illustrates an important feature: the total impedance of the circuit is a *combination* of R and X_L, and Z in this circuit leads current slightly because of the resulting value of X_L. In a predominantly capacitive circuit, the value of X_C would predominate and the circuit impedance (Z) would lag the current. This point is very important and is discussed again in question E.3.7, below.

The combination of X_L and R, or X_C and R, may be found by using the formula $Z = \sqrt{X^2 + R^2}$, where X^2 represents the value of $X_L - X_C$ or $X_C - X_L$. In the example shown in Fig. 38(c), the impedance (Z) is equal to 10 Ω. Z is determined as follows:

$$\begin{aligned} Z &= \sqrt{X^2 + R^2} = \sqrt{X_L{}^2 + R^2} \\ &= \sqrt{6^2 + 8^2} = \sqrt{36 + 64} \\ &= \sqrt{100} = 10\ \Omega \end{aligned}$$

E.3.3(d) How do like reactances combine in series?

Answer: Like reactances in series are added to determine the total reactance.

Discussion: In a series circuit, the current flows through each reactive component in turn. The total reactance of the series is found by adding the reactance of each component. This rule holds true for series-connected inductive reactances or for series-connected capacitive reactances and can be expressed as $X_{total} = X_1 + X_2 + X_3$, etc.

E.3.3(e) How do like reactances combine in parallel?

Answer: The combined reactance of like reactances in parallel is less than that of the lowest value reactance.

Discussion: Current flow is divided because of the parallel paths offered by each reactance. The combined value of parallel-connected like reactances is found by using either of the following formulas:

$$X_{total} = \frac{X_1 \times X_2}{X_1 + X_2} \quad \text{or} \quad X_{total} = \frac{1}{\frac{1}{X_1} + \frac{1}{X_2} + \frac{1}{X_3} \text{ (etc.)}}$$

E.3.4(a) What is Ohm's Law?

Answer: Ohm's law expresses the direct relationship between voltage, current, and resistance in a circuit. It is an algebraic equation stated as I (current in amperes) equals E (voltage in volts) divided by R (resistance in ohms), or, $I = E/R$. Algebraically this can be transposed to read $R = E/I$ and $E = I/R$. Thus, knowing any two values, the third value can be found. (See the discussion of Ohm's law under F.2.2 in Chapter 2.)

E.3.4(b) How does Ohm's Law relate to resistive and reactive impedances?

Answer: Ohm's law applies equally to resistive circuits and reactive impedance circuits.

Discussion: In a circuit containing reactive components, combined reactance (X) can replace R in the equation for Ohm's law. The symbol X is used for reactive impedance since the symbol Z usually indicates total impedance, reactive and resistive. The Ohm's law equation for reactive impedance circuits is: $I = EX$, $X = EI$, and $E = I \times X$. For a circuit in which the impedance Z is known, Z can be substituted for X or R.

E.3.5 What is impedance matching and when is it important?

Answer: Impedance matching occurs when a source of power and the load to which the power is being delivered have the same (matching) values of impedance. It is important in obtaining maximum transfer of power.

Discussion: A matching source will deliver maximum power to a load regardless of whether the circuit is ac-operated, such as the transmitter final amplifier to a transmission line, or dc-operated, such as a battery to a resistor. The important factor is that the impedance of both source and load match as closely as possible. A perfect match is the most desirable.

(For an explanation of how matching provides maximum power output, refer to question F.2.3 in Chapter 2.)

E.3.6(a) What is a decibel?

Answer: A decibel represents one-tenth of a unit of power ratio measurement. The *bel* is a logarithmic unit of power ratio; *deci* stands for one-tenth.

Discussion: The decibel, as used for power or voltage calculations, is the logarithmic value of the power ratio. It is expressed as $\text{dB} = 10 \log_{10} (P2/P1)$, where $P1$ is input power and $P2$ is output power. Since common base 10 logarithms are used, the equation is most often expressed as $\text{dB} = 10 \log (P2/P1)$, or for voltage, $\text{dB} = 20 \log (V2/V1)$. When the ratio is given for an increase in power, or voltage, the decibel has a positive value; for example, a 6-dB gain. When the ratio is given for a decrease in power, the decibel has a negative value; for example, a −3-dB loss.

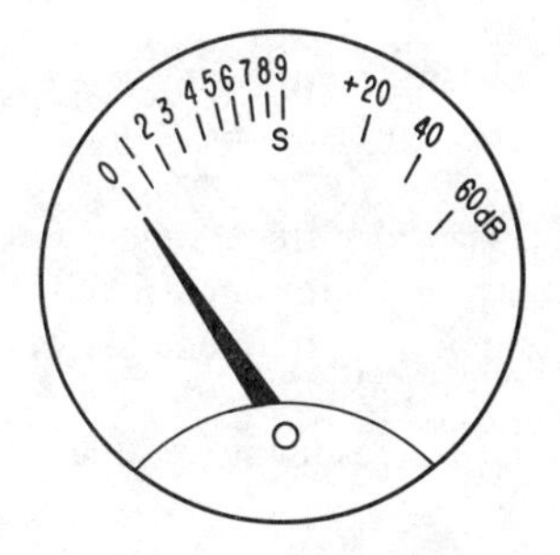

Fig. 39. Signal-strength meter (S-meter) calibrated in S units and decibel (dB) values above S9.

E.3.6(b) What is an "S" unit?

Answer: An *S* unit is a receiver signal level unit of measurement ranging from 0 to 9. Each S unit usually has a value of 6dB.

Discussion: Most amateur receivers use a front panel signal strength meter, known as an S-meter. As shown in Fig. 39, the S-meter most often has calibrations ranging from S0 to S9. Values above S9 are calibrated in dB. Should a 20-dB signal be received, the S-meter would read "20 dB over S9." Each S unit value is usually rated at 6 dB; thus, an S9 signal indicates a total signal strength of 54 dB.

E.3.7(a) What are the distinguishing features between series- and parallel-resonant circuits?

Answer: A series-resonant circuit offers maximum current flow and minimum impedance to the generator source. A parallel-resonant circuit offers minimum line current (maximum current flows in the tank circuit) and maximum impedance to the generator source.

Discussion: It is necessary to discuss resonance before discussing the distinguishing features of series- and parallel-resonant circuits. As noted in E.3.3(c), the actions of X_L and X_C cancel each other. And, as noted in the discussion of E.3.3(a) and (b), in the formulas for finding the values of X_L and X_C, frequency (f) was the one variable value. If the values of inductance and capacitance are held fixed, the different values of frequency applied to the circuit containing the inductor or capacitor would vary the value of X_L or X_C. Note that the formula for X_C expresses an inverse proportion, which means that as the frequency applied to the capacitor *increases*, the value of X_C *decreases*. The formula used to determine the inductive reactance (X_L) expresses a direct proportion. An *increase* in the frequency applied to an inductor causes an *increase* in X_L.

An inductor and capacitor in a series-resonant circuit is shown in Fig. 40. The value of R in the circuit represents the resistance offered by the turns of wire of the inductor. As noted in the discussion of E.3.3(c), the reactance of a

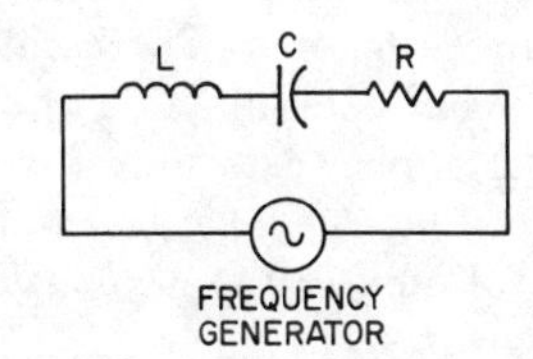

Fig. 40. Series-resonant circuit.

series circuit containing X_L and X_C is determined by subtracting the smaller value from the larger; the remaining value (X_L or X_C) determines the reactance of the circuit.

Since X_L increases with frequency and X_C decreases with frequency, there is *one* crossover frequency for any series combination of inductor and capacitor in which the values of X_L and X_C will be exactly equal. This frequency is the *resonant* frequency.

The following formula is used to find the value of the resonant frequency:

$$f = \frac{1}{2\pi\sqrt{LC}}$$

where L is in henries and C is in farads.

At resonance in a series circuit in which X_L and X_C cancel each other, the only opposition to current flow from the generator frequency source is the resistance. When the value of R is low, the current flow is high. Thus, the distinguishing feature of the series resonant circuit is the minimum impedance offered at resonance, which results in a high value of current flow.

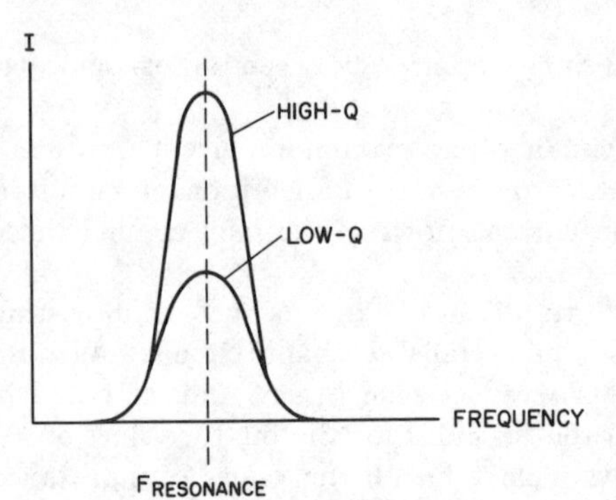

Fig. 41. Resonant frequency curves.

To illustrate the response of a resonant circuit, a curve is drawn which plots the value of current versus the frequency. Figure 41 shows that the high-Q curve rises steeply as the frequency nears the resonant value and then falls off steeply after the resonant value, with maximum current flowing at the resonant frequency. The low-Q resonant curve in Fig. 41 shows a lower value of current flow at resonance, indicating that at resonance the resistance of the circuit offers an increased opposition to current flow due to a higher value of circuit resistance.

The formula $Q = X/R$ will help determine the quality (Q) of a resonant circuit. In circuits below 30 MHz, the resistance offered by the leads of a capacitor can be ignored. The comparison of inductive reactance to coil resistance is the major determination of circuit Q for these frequencies. A high-Q coil is an example of one that offers at resonance an X_L of 270 Ω and a resistance of 3 Ω for a Q of 90. A similar coil wound to offer less inductance might provide a value of X_L of only 150 Ω, while still offering a dc resistance of 3 Ω to have a Q of 50.

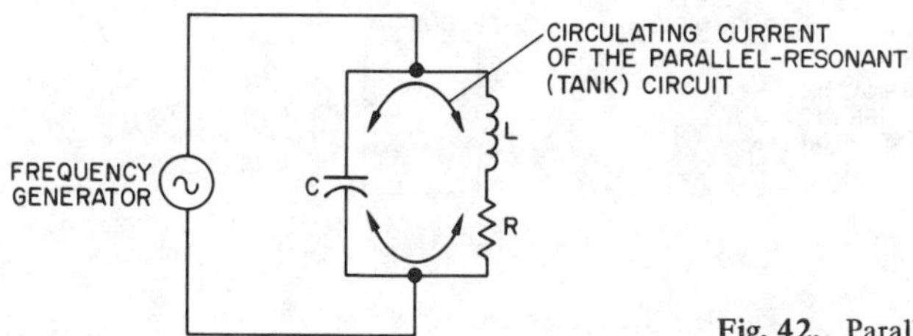

Fig. 42. Parallel-resonant circuit.

In the parallel-resonant circuit of Fig. 42, the same conditions for resonance exist in that at the resonant frequency the values of X_L and X_C are equal. The action, however, differs greatly from that of a series circuit. At resonance, the initial current flow charges the capacitor after a period of time determined by the frequency of resonance. The capacitor discharges and the current flow of the discharge builds a magnetic field around the inductor. After the same period of time, the capacitor, being fully discharged, can no longer maintain the magnetic field of the inductor, causing the magnetic field to collapse. A current then develops in the coil and flows in the opposite direction. This current flows into the capacitor to recharge the capacitor in the polarity *opposite* that of its original charge. The double-headed arrows indicate the back-and-forth current flow within the LC circuit of Fig. 42. Because the energy stays within the LC circuit, traveling back-and-forth at a rate determined by the frequency of resonance, the LC circuit is often referred to as a *tank* circuit.

The distinguishing features of a parallel-resonant circuit are: At resonance, the current within the tank circuit is maximum. The only current flow from the generator is the small amount required to make up the resistance losses of the coil winding. The low value of line current from the generator to the parallel-resonant circuit indicates that at resonance the tank circuit offers a high impedance.

The resonant curves showing the Q of a parallel-resonant circuit are similar to those of a series-resonant circuit; the difference is that the current indicated in the vertical (Y) axis is that of the circulating tank current.

E.3.7(b) How is the resonant frequency of series- and parallel-resonant circuits determined?

Answer: The resonant frequency is determined by the values of L and C used. It is found by using formula,

$$f = \frac{1}{2\pi\sqrt{LC}}$$

(The determination of resonant frequency is discussed in E.3.7(a) above.)

E.3.7(c) What is the Q of a series-resonant circuit? Of a parallel-resonant circuit?

Answer: The Q of a series- or parallel-resonant circuit is determined by dividing the circuit reactance by the circuit resistance: $Q = X/R$. (The determination of Q is discussed in E.3.7(a) above.)

E.3.8(a) What is the operating principle of a transformer?

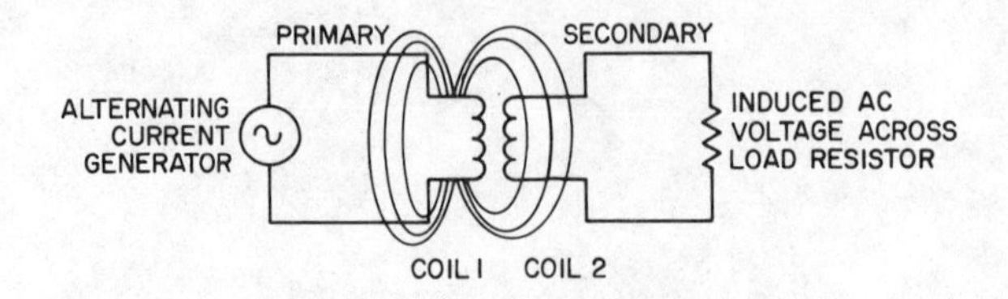

Fig. 43. Basic transformer action.

Answer: The expanding and contracting magnetic field of a coil carrying alternating current will induce an ac voltage in a second coil (placed nearby) when the field cuts across it.

Discussion: Application of ac to a coil of wire produces a magnetic field around the coil, which expands and contracts with the alternating current. A transformer is created by placing a second coil beside the first coil, causing the expanding and contracting magnetic field of the first coil to cut through the turns of wire of the second coil (referred to as mutual induction) (Fig. 43). The expanding and contracting magnetic field of coil 1 induces a voltage in coil 2, which is applied to whatever load is placed across the turns of coil 2. Coil 1 is referred to as the *primary* winding of the transformer because the originating voltage is applied to it, and coil 2 is referred to as the *secondary* winding.

E.3.8(b) How does a transformer provide a desired voltage?

Answer: Voltage can be stepped up or down as desired by adjusting the ratio of the number of turns of the secondary winding to the number of turns of the primary winding.

Discussion: The total voltage applied to the primary winding will be developed in the secondary if the number of turns in each winding is the same, if there are no losses, and if the coupling between the windings is perfect. For example, 50 volts applied to the primary will develop 50 V across the secondary winding if there are 100 turns in each winding.

If it is desirable to step up input voltage from 50 to 100 V, the secondary coil can be wound with 200 turns of wire to provide a 100 turns to 200 turns ratio (1:2), referred to more often as a *2:1 step-up ratio* (Fig. 44).

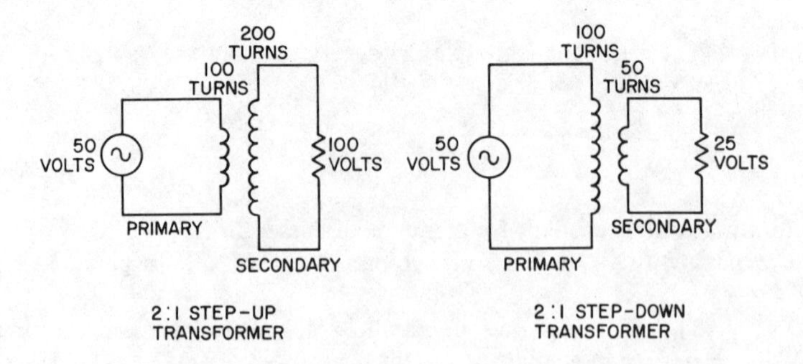

Fig. 44. Using different turns ratios to provide step-up and step-down transformers.

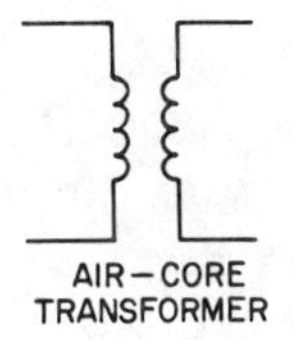

Fig. 45. Symbols for air-core and iron-core transformers.

When a one-half reduction in voltage to the secondary is desired, the number of turns in the secondary winding is one-half that of the primary or, in the example shown, 50 turns. This would be a *2:1 step-down ratio,* providing a 25-V output.

The losses briefly mentioned above result from the air space between windings. Transformers with air spacing between windings are most common in higher radio frequency operations. Special powdered iron cores are used between windings to decrease losses. The losses in air coupling are high for lower audio or power-line frequencies. To cut down on these losses laminated layers of iron are used between the windings. Lines are drawn between the windings to indicate an iron-core transformer (Fig. 45).

Transformer losses are due mainly to wire resistance losses, to heating of the core, to leakage reactance where the magnetic field does not cut through the total winding. In iron core transformers, hysteresis and eddy current losses result from the loss of energy required to continually reverse the magnetic polarity of the core material. The efficiency of a transformer would be its input, less its losses as developed in the output, expressed in percentage. For example, if a transformer primary consumed 100 W of power, and the losses were such that only 75 W were available at the output of the secondary, the 25-W loss would be noted by saying that the transformer was 75% efficient.

E.3.8(c) How does a transformer provide a desired impedance?

Answer: The impedance matching ability of a transformer is similar to its ability to step a voltage value up or down. A winding with many turns has a high reactive impedance; a winding with few turns has a low reactive impedance. Therefore, a primary winding with many turns and a secondary winding with few turns offers a step-down impedance ratio. A primary winding with fewer turns than the secondary winding offers a step-up impedance ratio.

Discussion: The impedance matching ability of a transformer is related to the ratio of the number of turns of the primary to the number of turns of the secondary. The formula is:

$$\sqrt{\frac{Z_{sec}}{Z_{pri}}} = \frac{N_{sec}}{N_{pri}}$$

where N = Number of turns.

Although many factors are involved in determining the exact impedance values, transformers such as those shown in Fig. 44 offer the same type of impedance match as they offer for a voltage match. A step-up transformer offers a step-up in both impedance and voltage, and a step-down transformer offers a step-down in both impedance and voltage.

F. Practical Circuits

F.3.1 Why do circuits oscillate?

Answer: Circuits oscillate when a sufficient portion of the output of an amplifier is fed back to the input of the amplifier. The feedback must be in phase for oscillation to occur.

Discussion: Circuit oscillation can be both desired and undesired. As noted above, feeding a portion of the output signal back to the input circuit causes the amplifier to reamplify the fed-back signal. In-phase feedback means that the output signal which is fed back has the same polarity as the input signal. In-phase feedback is also called positive feedback. Positive feedback increases the signal strength at the input to the point where the input signal is excessive, causing the amplifier to develop self-sustaining oscillation. Because the effect is self-sustaining, removing any externally applied signal from the input will not stop the oscillation as long as the positive feedback remains.

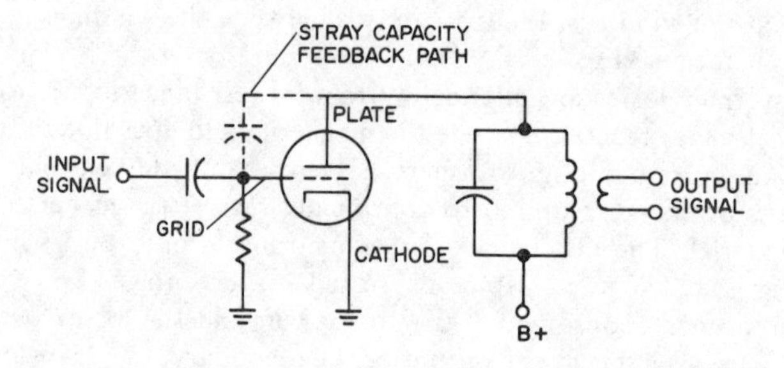

Fig. 46. Undesired feedback path between tube elements.

Undesired oscillation usually occurs as a result of stray capacitance in circuit wiring or interelement stray capacitances in a tube or transistor (Fig. 46). A portion of the amplified output at the plate can be coupled back through stray capacity to the grid input circuit to have an amplifier break into undesired oscillation.

Desired oscillation in a tube and transistor version of a Colpitts-type oscillator appears in Fig. 47. The frequency-determining tank circuit is in the grid, or input circuit, of the tube-type oscillator and in the collector, or output circuit, of the transistor-type oscillator. Operation is identical for each type. In the tube-type circuit, a portion of the amplified signal at the output (plate) circuit is coupled through a capacitor to the tuned circuit at the input (grid) to provide positive (in-phase) feedback that will sustain oscillation. In the transistor-type circuit, feedback is from the output (collector) circuit to the input (base) circuit to sustain oscillation.

In the tube circuit, the inverted output signal is applied to the bottom end of the grid circuit tank coil which is 180° out of phase with the top end of the coil. With 180° inversion in the tube amplifier circuit and 180° inversion in the coil, the total inversion is 360°, or in-phase feedback. In the transistor circuit, the output signal fed back to the base is amplified and inverted 180°. This

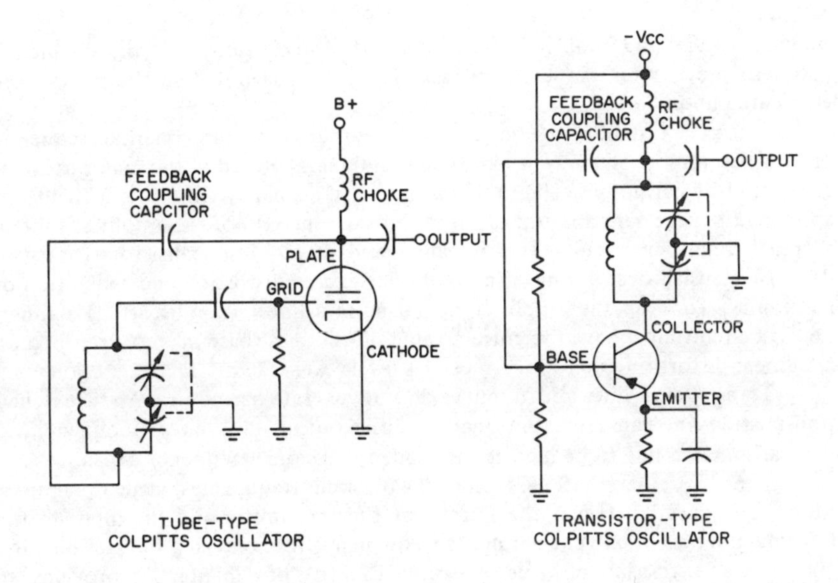

Fig. 47. Tube- and transistor-type oscillator circuits.

inverted signal is applied to the bottom of the collector circuit tank coil which is 180° out of phase with the top of the coil. The combined inversion of the transistor and coil taps creates a 360° inversion, or in-phase feedback.

Note that in the transistor oscillator circuit base bias is fixed by using a voltage divider from $-V_{cc}$ to ground. The tube oscillator circuit uses self-bias in the form of grid-leak bias. In CW operation, closing the key can cause the output signal to "load" the oscillator circuit, causing it to change frequency. Having the oscillator output applied to a buffer amplifier will minimize frequency drift due to loading.

F.3.2 What are some ways of minimizing harmonic generation in frequency doublers, vacuum tube amplifiers, transmission lines, and antennas?

Answer: For frequency doublers and amplifiers, all resonant circuits must be properly tuned. The minimum necessary value of grid drive should be used. All amplifiers should be linear and should use the correct value of grid bias. Transmission lines should be of matching impedance to the output circuit of the transmitter. To reduce harmonic output, a transmatch or low-pass filter can be used between the output of the transmitter and the transmission line, and to prevent radiation of harmonic frequencies, a nonresonant antenna can be used.

Discussion: Harmonic frequencies are multiples of the original or fundamental frequency. A second harmonic is twice the frequency of the fundamental; a third harmonic, triple, and so on. Tuned circuits are resonant at the fundamental and harmonic frequencies and generate harmonic frequencies if the circuit is specifically designed for this purpose. For example, a frequency doubler with an input signal of 15MHz can have an output circuit tuned to 30 MHz. For circuits designed to be operated at a specific frequency, harmonic frequencies are undesirable, therefore, specific precautions can and should be taken to prevent undesired harmonics. The methods noted in the answer are elaborated upon below.

Incorrect tuning of a resonant circuit either at the fundamental frequency or at a harmonic frequency, such as in a doubler, is probably the major cause of generation of harmonics. The tuning problem stems partly from the difficulty of applying the necessary amount of input signal to the grid of the amplifier tube in a tuned amplifier circuit. This is called grid drive. (For transistors it is base drive.) Extensive overdriving of an amplifier causes distortion and generation of harmonics. To keep the amplifier operating in a linear (nondistorted) manner, the grid circuit must have the correct value of grid bias (base bias for transistors). Nonlinear distortion due to incorrect bias levels causes harmonic generation.

Transmission lines and antennas do not generate harmonics. A mismatched transmission line can cause the final amplifier output tank circuit connected to the transmission line to be incorrectly loaded; harmonic frequencies result.

Tuned circuits that pass only the desired frequencies, called bandpass filters, are placed between the output of the transmitter and the input to the transmission line to prevent harmonic frequencies from entering the transmission line. Using a transmatch (see question I.2.7(b) of Chapter 2) provides an impedance match. Also, a transmatch provides a tuned circuit that passes only the desired band of frequencies, thus minimizing harmonic radiation. Another popular way to eliminate harmonics is to use a low-pass filter between the transmitter and transmission line. A low-pass filter, as the name implies, passes the lower frequencies but blocks, or attenuates, the higher harmonic frequencies.

A nonresonant antenna, that is, an antenna that will be resonant at the fundamental frequency but nonresonant at harmonic frequencies can minimize radiation of harmonic frequencies. Nonresonant antennas include the folded-dipole, a long-wire antenna terminated at the end through a resistor to ground, or a rhombic antenna whose the ends are terminated to ground through resistors.

F.3.3(a) What are the main classes of vacuum tube amplifier operation?

Answer: The main classes are Class A, B, and C.

F.3.3(b) For what use is each class best suited?

Answer: Class A is best suited for linear amplification with minimum distortion. Class B is used for increased power output and can also be used as a linear amplifier. Class C amplifiers are used for amplification of radio frequency signals.

Discussion: Figure 48 illustrates Class A, B, and C operation. Class A amplifiers are usually referred to as *linear* amplifiers since there is no distortion in the output. This lack of distortion results from limiting the input signal swing to the linear portion of the curve. The output signal is a full reproduction of the input signal. Because plate current is always flowing, even with no signal input, efficiency of the Class A amplifier is very low, approximately 20%.

Class B amplifiers operating at or near cutoff are also considered linear amplifiers despite the slight distortion due to nonlinearity of the grid-voltage plate-current curve near the cutoff point. Class B can be used as an rf amplifier but is often used in a push-pull configuration as an audio modulation amplifier. Because plate current flows during only one-half of the cycle efficiency is close to 50%.

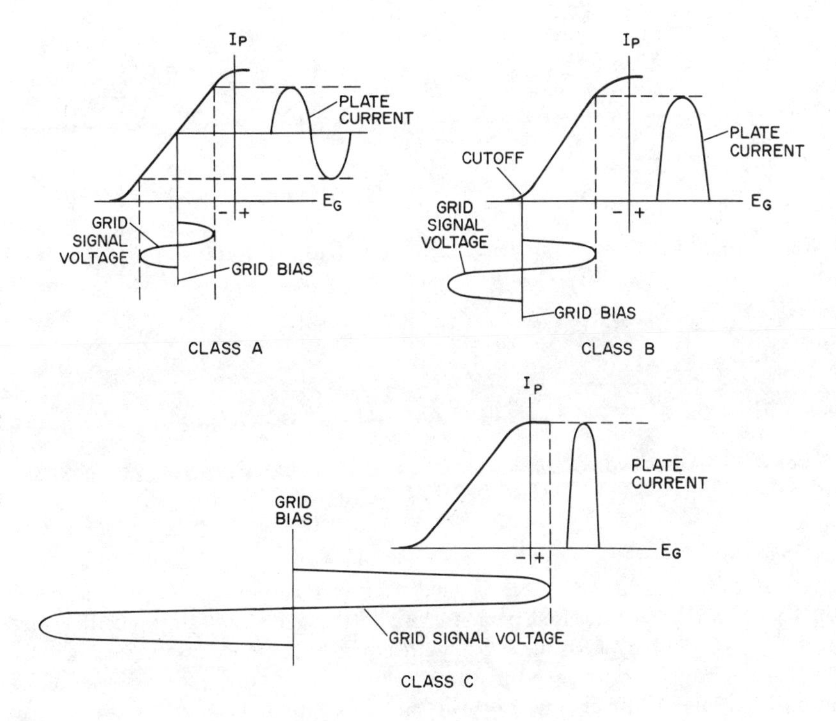

Fig. 48. Class A, B, and C operation.

Class C amplifiers are not linear amplifiers; they are *pulse* amplifiers. The tube conducts heavily for a short period of time to supply a pulse of plate current to a resonant LC circuit (tank circuit) used as the plate load. The grid signal swing and bias value is usually set so that the tip of the signal swing just biases the grid positive sufficiently to just barely overdrive the tube (linearity is of no concern). Plate current efficiency can go as high as 80%. When overdriving the grid circuit, the grid goes positive and draws current. This grid-voltage grid-current represents power that must come from the preceding stage that is driving the grid circuit. To power the grid circuit the preceding amplifier must be a current amplifier (driver) stage.

F.3.3(c) What are the typical efficiencies associated with each (class of amplifier)?

Answer: Class A is approximately 20% efficient; Class B, approximately 50%, and Class C, approximately 80%. (See the discussion of F.3.3(b) above.)

F.3.4(a) What is neutralization and how does it contribute to proper amplifier operation?

Answer:·Neutralization is the prevention of undesired positive feedback; it is used to prevent an amplifier from breaking into self-oscillation.

Discussion: Rf amplifiers present a problem in that a positive feedback path exists through the grid-to-plate interelectrode capacitance of the amplifier tube. This feedback can cause the amplifier to break into oscillation. Where the

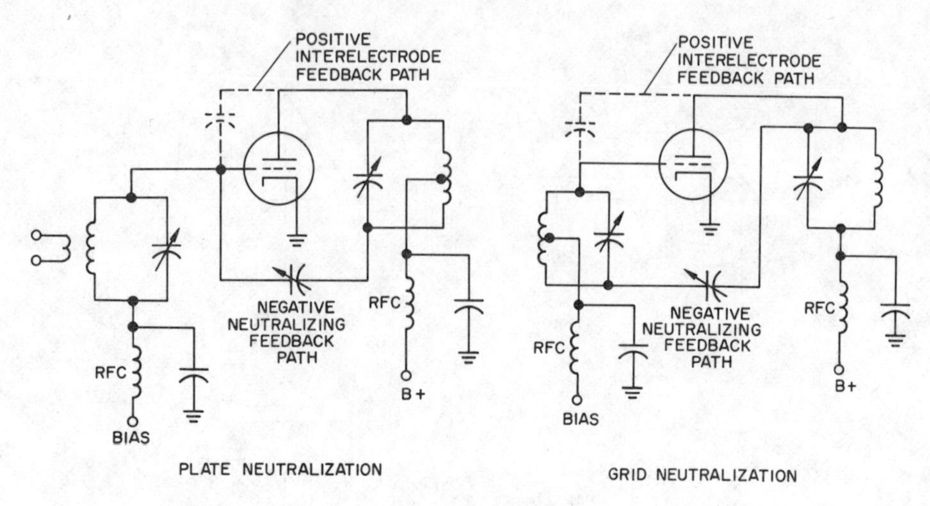

Fig. 49. Neutralization methods.

interelectrode capacitance value is small, the feedback is limited, but it can be sufficient to create undesired parasitic oscillation. This undesired positive feedback is prevented by neutralizing it with equal but opposite values of negative feedback. The neutralizing points are feedback from the plate to grid of the amplifier. This feedback, negative in sign but equal in value to the positive feedback, neutralizes the undesired positive feedback, thus preventing oscillation.

Figure 49 shows the two most common forms of neutralization, plate neutralization and grid neutralization. In plate neutralization, the plate tank coil is center-tapped and bypassed to allow either end of the tank coil to be tapped and a portion of the tank coil voltage to be capacity-coupled to the grid circuit. Tapping from the bottom of the coil provides out-of-phase voltage to neutralize the in-phase feedback through the plate to grid interelectrode capacitance. In grid neutralization, bias is applied to the grid through a center tap on the grid circuit tank coil permitting feedback to either end of the grid tank coil. Neutralizing feedback is taken from the top of the plate tank coil and applied to the bottom of the grid tank coil.

F.3.4(b) What procedures should be followed to neutralize an rf amplifier properly?

Answer: The following procedures should be followed:

1. Remove the plate voltage (and the screen voltage in the case of a tetrode) of the tube in the stage being neutralized but leave the filament power applied.
2. With the power on, tune the stage that feeds the stage being neutralized to the proper frequency.
3. Using an rf indicator, tune the input (grid) and output (plate) tank circuits of the stage being neutralized to resonance with the input frequency from the preceding stage.

4. Adjust the neutralizing capacitor and retune the input and output tank circuits until there is minimum indication of rf in the plate tank circuit. Repeat the procedure until the final setting of the neutralizing capacitor is obtained.

Discussion: An rf indicator can consist of a low-power bulb such as a flashlight or neon bulb connected to a small loop or a loop of a few turns of insulated wire. The wire loop is held close to the tank coil to induce a voltage in the loop, causing the light to glow. The brilliance of the bulb is the indicator.

Adjusting the neutralizing capacitor and retuning the input and output tank circuits continue until a minimum value of rf indication appears on the bulb. (The preferred value is zero which, however, is often an unattainable value.)

F.3.5 Compare the operating characteristics of grounded-grid, grounded-cathode, and grounded-plate amplifiers.

Answer: Grounded-grid amplifiers have low interelectrode capacitance, making them useful at higher frequencies. No neutralization is required for amplification of radio frequencies. Such amplifiers offer low input impedance and high output impedance. The output signal is in phase with the input signal.

Grounded-cathode amplifiers are conventional amplifiers, and require neutralization for amplification of radio frequencies. They offer high input and high output impedance. The output signal is 180° out-of-phase with the input signal.

Grounded-plate amplifiers are commonly called cathode-followers, and do not require neutralization for amplification of radio frequencies. They offer high input impedance and low output impedance. The output signal is in phase with the input signal. (Figure 50 illustrates the three basic types of amplifier circuits discussed here.)

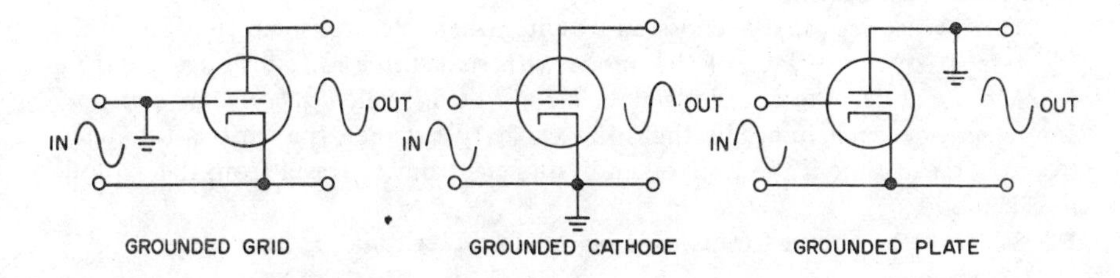

Fig. 50. Basic amplifier circuit configurations.

F.3.6 Why are resonant circuits used in rf amplifiers?

Answer: Resonant or tank circuits are used in both receiver and transmitter rf amplifiers because of their ability to select or tune to a specific frequency or to a specific band of frequencies.

Discussion: Resonant circuits have been discussed in questions E.3.7(a), (b), and (c), above. The frequency of resonance, f_0, is the specific frequency to which a resonant circuit is tuned.

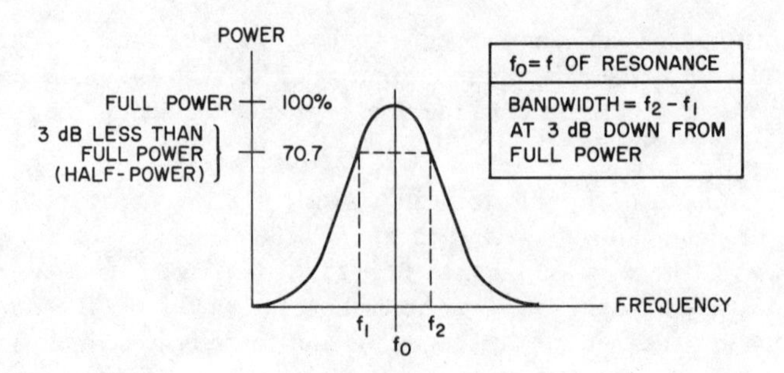

Fig. 51. Determining the bandwidth of a resonant circuit.

The band of frequencies passed by a tuned circuit, called the bandpass or bandwidth of a tuned circuit, is determined by measuring the band of frequencies that fall between the half-power points in the curve (Fig. 51). At 70.7% of the maximum power of the tuned circuit, a logarithmic value exists, which is 3 dB less than at the full power value. This value is known as "3 dB down" on the curve. Because power is determined by finding the current *squared* times the impedance, the true power value of 3 dB down, or 70.7% of the current, is half the value of full power. Thus, there are three identities for the point on the curve to determine bandwidth, all meaning the same. The identities are: 3 dB down, 70.7% of the current, and half-power point. Subtracting the lower frequency from the higher frequency at the half-power points gives the formula for bandwidth shown in Fig. 51.

F.3.7 In a vacuum tube what is meant by secondary emission; interelectrode capacitance; transit time?

Answer: Secondary emission results when electrons emitted from the cathode strike the surface of the anode with such force as to dislodge electrons from the anode, these dislodged electrons being called *secondary emission.* Interelectrode capacitance is the stray capacity between the various tube electrodes. Transit time is the time taken by the electrons to travel from the cathode to the anode.

F.3.8 Draw the circuit for cathode bias and grid-leak bias operation.

Answer:

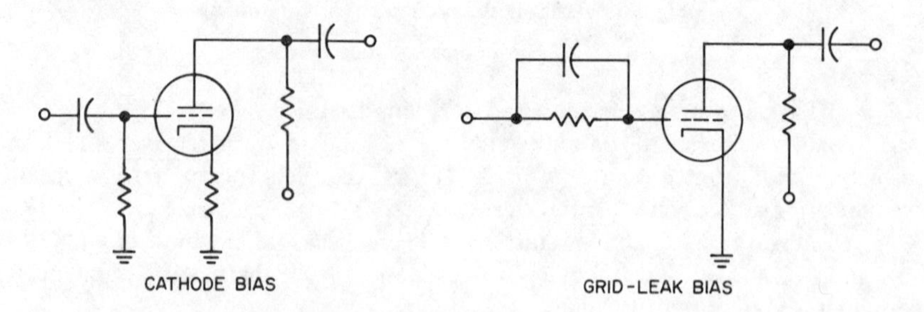

F.3.9(a) What is a bandpass filter?

Answer: A bandpass filter passes a specific band of frequencies while blocking the passage of all other frequencies.

F.3.9(b) What is a band-reject filter?

Answer: A band-reject filter will block a specific band of frequencies while passing all other frequencies.

F.3.9(c) What is a low-pass filter?

Answer: A filter circuit that passes all frequencies below a specific frequency and blocks all frequencies above the specific frequency.

F.3.9(d) What is a high-pass filter?

Answer: A filter circuit that passes all frequencies above a specific frequency and blocks all frequencies below the specific frequency.

F.3.10 What is meant by power supply voltage regulation?

Answer: To prevent the voltage output of a power supply from fluctuating the power supply voltage can be regulated using gaseous tube regulators, zener diodes, or a regulated "series" resistance, in which the series resistance is represented by a transistor or tube that is controlled to provide a varying resistance to the power supply to furnish a regulated output voltage.

F.3.11 Using transistors draw the circuits for direct coupling; capacitive coupling; impedance coupling; link coupling.

Answer:

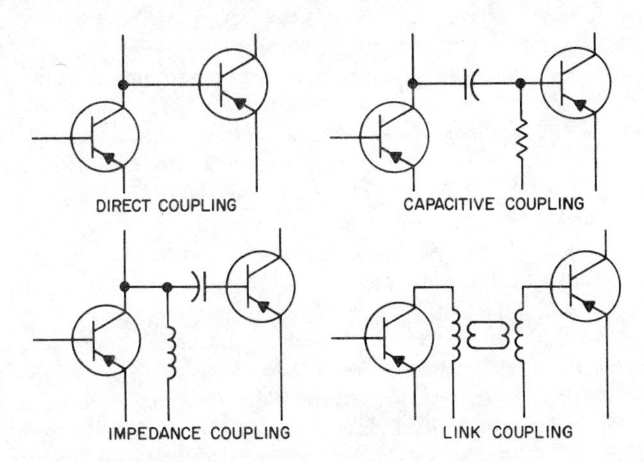

F.3.12(a) In a receiver, what is the purpose of the beat frequency oscillator?

Answer: CW signals cannot be detected using conventional diode detection. By beating the carrier code signals against another fixed frequency oscillator signal we can obtain a difference frequency signal that is in the audible range, approximately 1000 Hz, allowing us to copy the CW signals. The CW detecting oscillator is referred to as a *beat frequency oscillator.*

F.3.12(b) In a receiver, what is the purpose of automatic gain control (AGC) circuits?

Answer: The automatic gain control (AGC) circuit regulates the gain of the receiver keeping the output signal at a constant level.

F.3.13 What is an operational amplifier?

Answer: A direct-coupled, high-gain amplifier, usually an integrated circuit, that utilizes controlled feedback from the output to the input to regulate and maintain a specific gain level.

F.3.14(a) What is a logic OR circuit?

Answer: A special circuit, usually an integrated circuit, having two (or more) inputs in which there will be an output when any one *or* more of the inputs has correct signal levels applied.

F.3.14(b) What is a logic AND circuit?

Answer: A special circuit, usually an integrated circuit, having two (or more) inputs in which there will be an output only when input 1 *and* input 2 (or all inputs) have correct signal levels applied.

G. Circuit Components

G.3.1(a) What is the principle of a semiconductor diode?

Answer: A junction of P-type and N-type semiconductor material such as germanium or silicon forms a semiconductor diode. Applying dc voltage with the positive polarity applied to the P-type material and the negative polarity to the N-type material causes the semiconductor diode to be forward biased, permitting current to flow through the diode. Reversing the polarity of the applied voltage causes the diode to be reverse biased, thereby allowing no current to flow through the diode.

Discussion: Crystalline materials such as germanium or silicon are molecularly bound by the electrons in the outermost orbit of the atomic structure. Being bound in this manner, the pure germanium or silicon is an insulator since it has no free electrons for current flow. Adding an impurity such as antimony, which has more electrons in its orbit than that of the germanium or silicon, creates an impure germanium or silicon that has free excess electrons that have been provided by the impurity. These available electrons allow the germanium or silicon to conduct a current, though not as well as copper wire or metal conductors. The impure germanium or silicon material is known as a *semiconductor.* Because the added electrons give the material a negative charge the semiconductor material is called "N-type."

Indium is an impurity that has fewer electrons in its orbit than germanium or silicon. Adding this impurity to either germanium or silicon creates a material with a deficiency of electrons. A space, or hole, exists where an electron should be. When an electron becomes available, it moves in to fill the hole. However, in doing so it leaves an empty space, or hole, in its original location. Thus, hole current flow can be explained as either the electron moving to the hole or the hole moving to position of the displaced electron. Within the semiconductor material, the flow of holes is accepted as the flow of current. The flow of holes stops where the material is joined by the wire conductor. An electron enters the material from the wire as a hole meets the wire conductor. Hole conduction is restricted to P-type semiconductor material. Since the material containing the impurity indium can conduct current, although not too well, the impure germanium or silicon is labeled a semiconductor. Because the holes are positive charges, this semiconductor material is called "P-type."

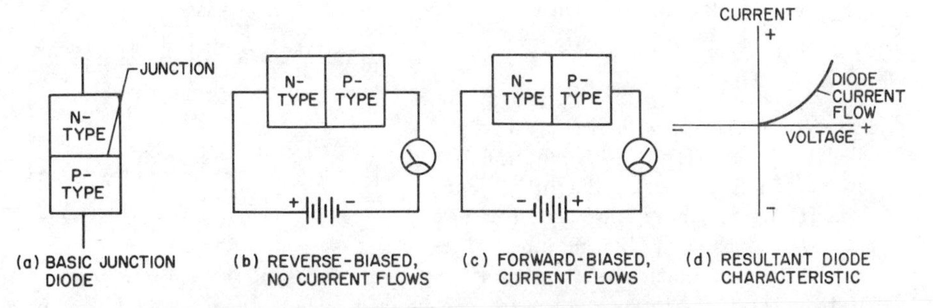

Fig. 52. A junction diode, reverse- and forward-biased.

Semiconductor diodes are made by special processes so that N- and P-type materials form alternate layers. The material can be cut so that a PN juction (Fig. 52a) is formed. The materials cannot merely be pressed together; they must have a junction with absolutely no division or space between the two materials.

Applying dc voltage with the polarities connected as shown in Fig. 52(b) is called reverse-bias and causes no current flow through the semiconductor diode because the positive terminal of the battery attracts and draws the electrons from the N-type material at the same time that the negative terminal of the battery draws the holes toward the connection of the wire to the P-type semiconductor (holes do not flow in the wire). This action is almost instantaneous, and after it has occurred, no current flows in the circuit. During reverse bias the junction appears as a capacitor. This junction capacitance must be taken into consideration when the diode is used in high frequency circuits. Applying a dc voltage with the polarities shown in Fig. 52(c) is called forward-bias and causes current to flow through the semiconductor diode because the electrons of the N-type material are both attracted and pushed to the junction where they fill the holes of the P-type material. The attraction is from the positive charge of the holes and the push is from the repelling action of the negative terminal of the battery (negative electrons being repelled by the like charge of the negative battery terminal). At the same time, the positive holes are being attracted to the junction by the electrons in the N-type material and the repelling action of the positive terminal of the battery. The electrons flowing from the negative terminal of the battery to the N-type material and from the P-type material to the positive terminal of the battery provide an electron current flow in the circuit. The resulting current flow is the characteristic curve of a diode (Fig. 52(d)).

G.3.1(b) What is the principle of a transistor?

Answer: Two junctions of N- and P-type semiconductor material that form either an N-P-N or P-N-P combination make up a transistor. The outer semiconductor materials are referred to as the collector and emitter and the center semiconductor material is a very thin layer called the base. The emitter-to-base junction is forward biased and the collector-to-base junction is reverse biased. The majority of the current carriers (holes or electrons) at the forward-biased low resistance emitter-to-base junction flows through the thin base region to control the reverse-biased high resistance collector-to-base current flow.

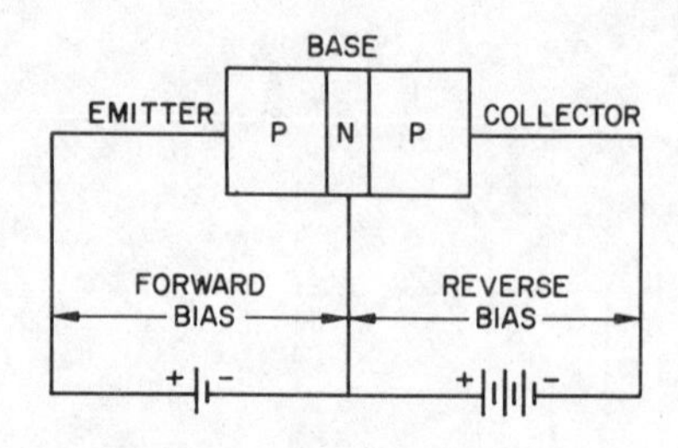

Fig. 53. PNP transistor circuit.

Discussion: Figure 53 illustrates a PNP-type transistor with forward bias applied to the emitter-base PN junction and reverse bias to the base-collector NP junction. (G.3.1(a), above, discusses forward and reverse biasing of NP junctions.)

With the emitter-base forward biased, the holes in the emitter and the electrons in the base move toward the junction where they would normally combine to create a complete circuit to have current flow. However, because of the very thin base and the attraction of the higher negative voltage at the collector, the majority (about 95%) of the holes from the emitter material will diffuse (pass) through the thin base and continue through the collector as hole current from emitter-to-collector. This results in approximately 95% of the emitter current flowing through to the collector and only 5% returning through the base to the emitter-base bias battery.

As a result, any change in current flow in the emitter-base circuit causes a corresponding change in current flow in the base-collector circuit. The emitter-base circuit is forward biased and offers low resistance; the base-collector circuit is reverse biased and offers high resistance. Current in the base circuit multiplied by the high value of resistance results in an $I \times R$ gain. The name transistor originated from the transfer of resistance from low (emitter-base) to high (base-collector), or a *trans*fer-res*istor*.

Circuit operation for an NPN transistor is identical to that of a PNP transistor except that the battery polarities are reversed and carrier current in the emitter and collector material is electrons.

G.3.1(c) Compare the elements of a solid-state diode with those of a vacuum tube.

Answer: The P-type material of a solid-state diode corresponds to the anode of a vacuum tube diode; the N-type material of the solid-state diode corresponds to the cathode of the vacuum tube diode.

Discussion: Figure 54 illustrates a junction diode and the schematic symbols for a semiconductor diode and a vacuum tube diode. Note that the part of both the tube and semiconductor diode to which the positive voltage is applied is called the anode and the part to which the negative voltage is applied

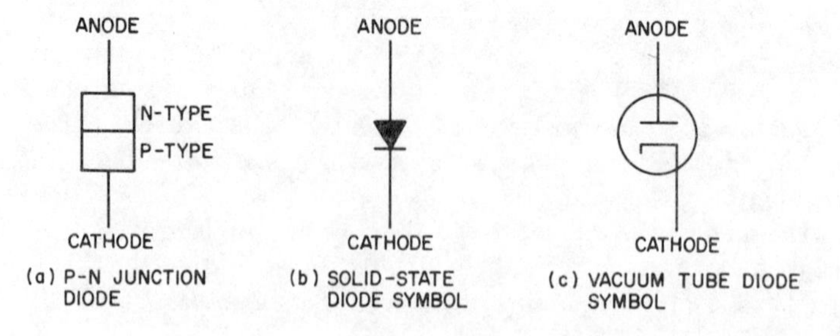

Fig. 54. Diode comparisons.

is called the cathode. Both diodes when used as rectifiers have the same operating terminals. In a vacuum tube diode, applying a positive voltage to the anode (usually called the *plate*) and a negative voltage to the cathode will cause the current to flow from the cathode to the anode.

G.3.1(d) Compare the elements of a transistor with those of a triode vacuum tube.

Answer: The emitter of a transistor is similar in operation to that of the cathode emitter in a vacuum tube. The base of the transistor is similar in operation to the grid of a vaccum tube. The collector of a transistor is similar in operation to that of the anode of a vacuum tube.

Discussion: Figure 55 illustrates the symbols used for NPN and PNP transistors and a triode vacuum tube. In both the vacuum tube and the transis-

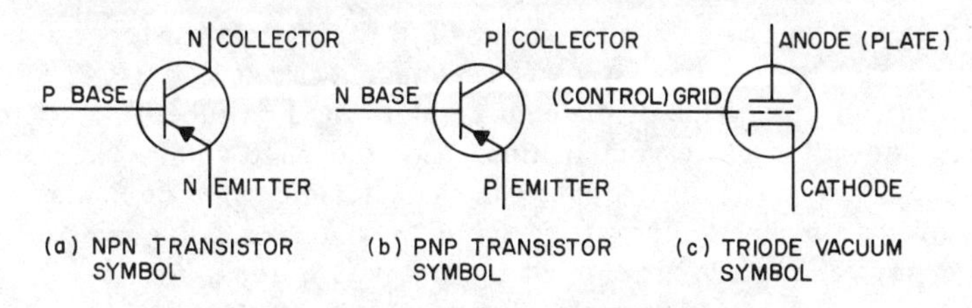

Fig. 55. Transistor and triode vacuum tube comparisons.

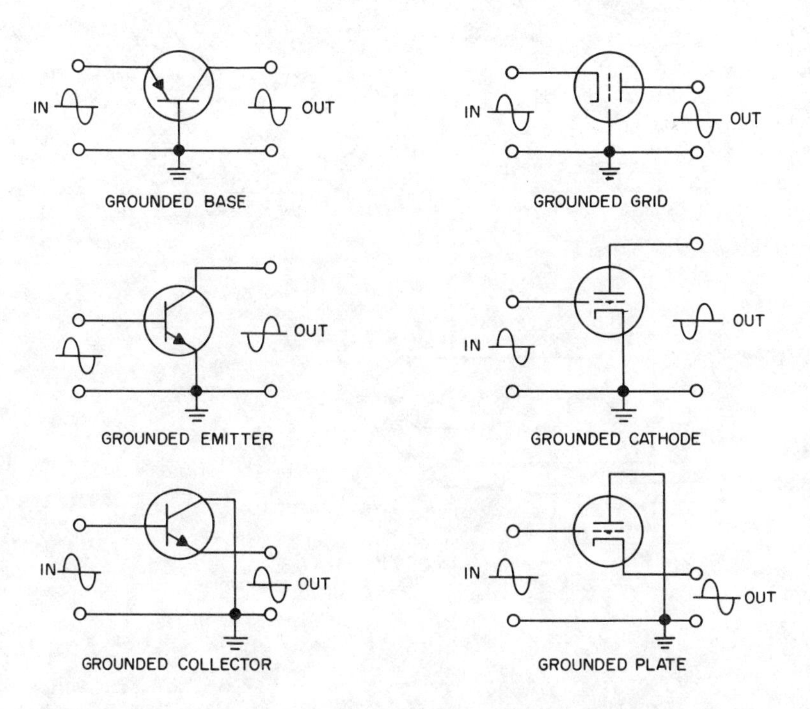

Fig. 56. Comparison of tube and transistor configurations.

tor, the original source of current flow is the cathode and emitter. The collector of the transistor is the point to which the major portion of the current flows within the transistor; the anode (plate) of the tube is the point to which all the current flows within the tube. The element that controls emitter-to-collector current flow is the transistor base and the element that controls cathode-to-anode current flow is the tube grid.

Figure 56 illustrates the similarity between the operation of tubes and transistors in the various configurations: grounded grid, plate, and cathode for tubes; and grounded base, collector, and emitter for transistors. Note that signal phase inversion takes place only in the grounded-cathode and grounded-emitter circuits.

G.3.1(e) What is the purpose of the characteristic curves of tubes and transistors?

Answer: The characteristic curves of tubes and transistors are used to determine how a tube or transistor will operate in a circuit.

Discussion: The value of plate current flowing in a tube is determined by the plate and grid voltage combination. With the plate voltage held constant varying the grid voltage will produce the characteristic curve of Fig. 57(a). The purpose of this curve is to illustrate the point at which increasing the grid voltage no longer increases plate current, plate current *saturation*. Also shown on the curve are the areas in which the grid voltage swing produces distorted output

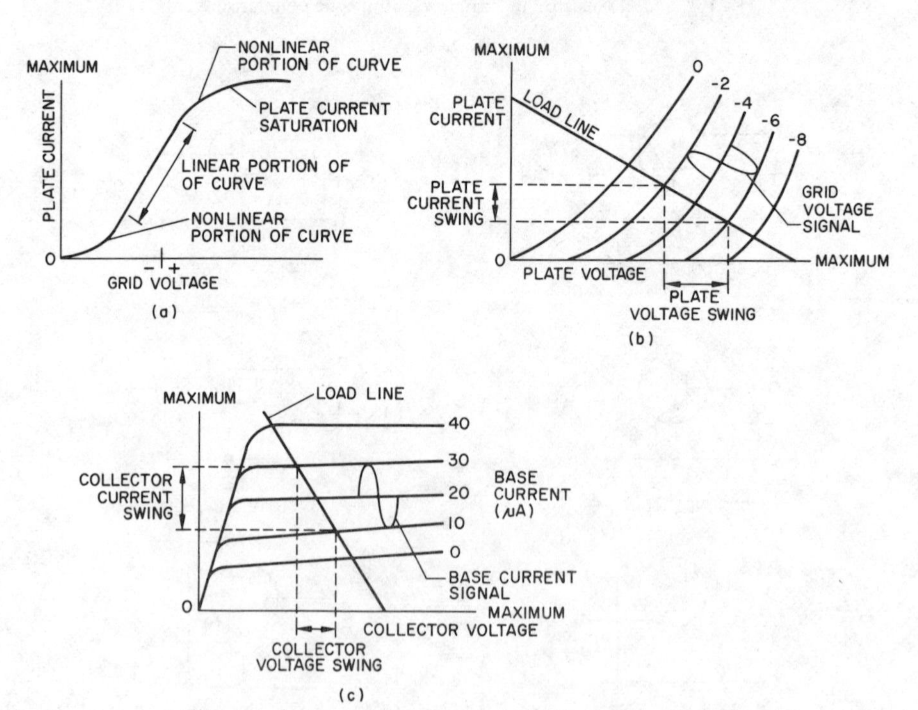

Fig. 57. Tube and transistor characteristic curves. (a) Plate current-grid voltage curve. (b) Plate current-plate voltage family of curves. (c) Collector current-collector voltage family of curves for a common-emitter amplifier.

signals because of the nonlinearity in the plate current curve. Of greater importance is the plate current-plate voltage curve in which the grid voltage is held constant and the plate voltage is varied, Fig. 57(b). By drawing the resulting curves for a group of grid bias voltages we have a *family* of curves. The family of characteristic curves can be used to determine the tube's plate resistance, amplification factor, or mutual conductance. When using the tube in a circuit the value of plate load resistance will determine the load line of the tube. This load line will show how the tube will operate by indicating what the plate voltage and plate current values will be for specific values of grid voltage, thus indicating how much the plate current and plate voltage will change with the change in grid voltage.

The most often used transistor characteristic curve is that of collector current versus collector voltage family of curves. These curves vary with each amplifier configuration, grounded base, emitter, or collector. The family of curves shown in Fig. 57(c) is for the more often used common emitter amplifier. The use of the family of curves is similar to that for a tube, with the major difference being that the input signal swing is noted in current values instead of voltage values.

G.3.2(a) What is the principle of an electrolytic capacitor?

Answer: An electrolytic capacitor is a polarized capacitor containing a very thin film dielectric between aluminum foil plates. Because of the thin film dielectric, electrolytic capacitors offer large values of capacity in a small physical area. Applying a dc current to the two aluminum foil plates separated by an oxide causes the dielectric to form and polarize so that current can be applied only to the capacitor in the same polarity as that used to form the dielectric.

Discussion: Electrolytic capacitors are constructed of very thin aluminum foil plates which are deliberately roughened (etched). The depressions and sides of the depressions caused by the foil roughness greatly increase the surface area. Between the foil plates is a very thin (thousandths of an inch) layer of oxide. Current flows through the oxide when a dc current is applied to the plates. In time, the current flow forms a thin film on the oxide next to the positively connected plate. This thin film, acting as an insulator, gradually reduces the current flow. When the film is completely formed, only a small "leakage" current flows.

Applying current in the reverse polarity causes the dielectric to function as a conductor. Therefore, the voltage applied to an electrolytic capacitor in a circuit must be dc voltage and must be applied as indicated on the terminals of the capacitor.

G.3.2(b) Why are electrolytic capacitors widely used in amateur radio equipment?

Answer: Electrolytic capacitors offer large values of capacity in small physical sizes at reasonable costs. (See the discussion of G.3.2(a) above.)

G.3.3(a) What is a toroidal inductor?

Answer: A toroidal inductor is a coil wound in a circular or doughnut shaped form.

Discussion: The wire is most often wound around a toroid-shaped core of powdered iron. Because of the circular shape, the electric field surrounding the coil is strongest at the hole or center of the coil and weakest around the outer edge. Therefore, the coil requires little or no shielding and offers little interference to adjacent components. Also, the strong field of a toroidal inductor increases the value of inductive reactance to provide increased *Q*. (Refer to E.3.7(a) above.)

G.3.3(b) What are the advantages and disadvantages of a toroidal inductor compared to an air-wound inductor?

Answer: The advantages of a toroidal inductor over an air-wound inductor are its higher *Q*, smaller physical size, and little or no shielding requirements. The disadvantage of a toroidal inductor is the difficulty in winding the coil. (See the discussion of G.3.3(a) above.)

G.3.4(a) What is meant by the *peak inverse voltage* rating of a rectifier diode?

Answer: The maximum or peak value of voltage that can be applied to a reverse-biased diode before the diode breaks down is the peak inverse voltage rating.

G.3.4(b) How can rectifier diodes connected in series be protected?

Answer: By placing equal value (equalizing) resistors across each diode to equalize the reverse voltage drop across each diode.

G.3.5(a) What is meant by the *piezoelectric effect?*

Answer: Certain crystalline materials when squeezed will develop a voltage; conversely, when a voltage is applied to the material the material bends, this is the piezoelectric effect. (See the discussion of G.2.1(d) in Chapter 2.)

G.3.5(b) What is the relationship of the thickness of the crystal to its operating frequency?

Answer: The thinner the crystal the higher the operating frequency.

H. Antennas and Transmission Lines

H.3.1(a) How is the approximate length of a half-wave dipole related to its resonant frequency?

Answer: When the length of the dipole antenna equals one-half the wavelength of the frequency being transmitted, the result is a resonant half-wave antenna.

Discussion: When an rf signal is applied to a resonant half-wavelength antenna such as that shown in Fig. 58(a), the signal travels outward exactly the

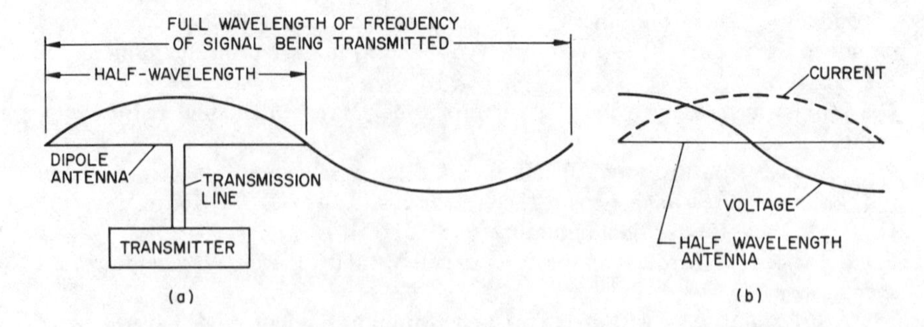

Fig. 58. (a) Resonant half-wavelength dipole antenna. (b) Current and voltage distribution over a resonant half-wave antenna.

full length of the antenna for the first half-cycle of the signal and then back to the transmission line feed point for the second half-cycle. Because the returning half-cycle meets the next half-cycle being transmitted, they combine to form a voltage and current distribution pattern (Fig. 58b). The voltage is maximum at the ends and current is maximum at the center. (The action of *end effect* antenna lengths, and the formula used to find the length of a half-wave dipole antenna are discussed in questions H.2.1 and H.2.2(b) of Chapter 2.)

H.3.1(b) How is the approximate length of a quarter-wave antenna related to its resonant frequency?

Answer: When the length of the antenna equals one-fourth of the wavelength of the frequency being transmitted, the result is a resonant quarter-wave antenna.

Discussion: Refer to the discussion of H.3.1(a) above. The formulas used to find the length of a half-wave dipole can be used to find the length of a quarter-wave antenna. The determined value of the length, however, must be halved because the quarter-wave is equal to one half a half-wave.

H.3.2 Compare the characteristics of a horizontal half-wave dipole with those of a quarter-wave ground plane antenna.

Answer: A horizontal half-wave dipole antenna operates in free space. Radiation is bidirectional (two directions) and the signal polarization is horizontal. The angle of radiation is normally low, but if the antenna is close to the ground, the angle of radiation increases. The impedance offered at the center of the antenna is about 72 Ω.

A quarter-wave ground plane antenna has the quarter-wave section radiating against the ground plane portion of the antenna. The ground plane acts as a mirror image to provide the other quarter-wave reflection; thus, the total antenna acts as a half-wave antenna. Radiation is omnidirectional (all directions), and the signal polarization is vertical. The angle of radiation is low. Impedance offered at the junction of the quarter-wave radiating element and the ground plane is lower, close to 30 Ω, than that of a dipole.

Discussion: Half-wave dipole antenna radiation is illustrated in Fig. 59. The top view of the antenna (Fig. 59a) shows that radiation is from the center out toward each side (bidirectional). The end view of Fig. 59(b) shows the low,

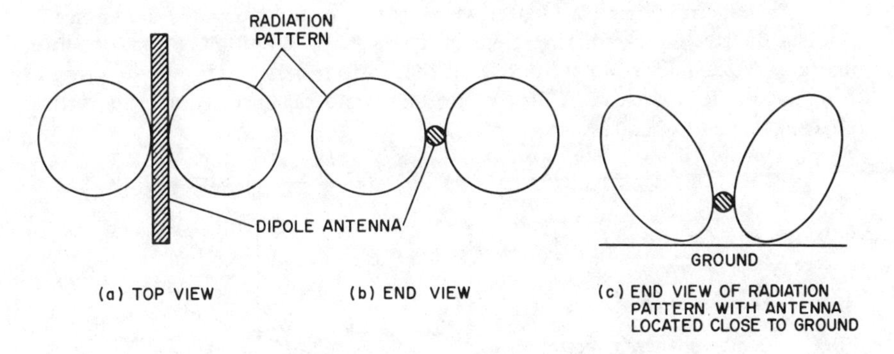

Fig. 59. Radiation pattern of a dipole antenna.

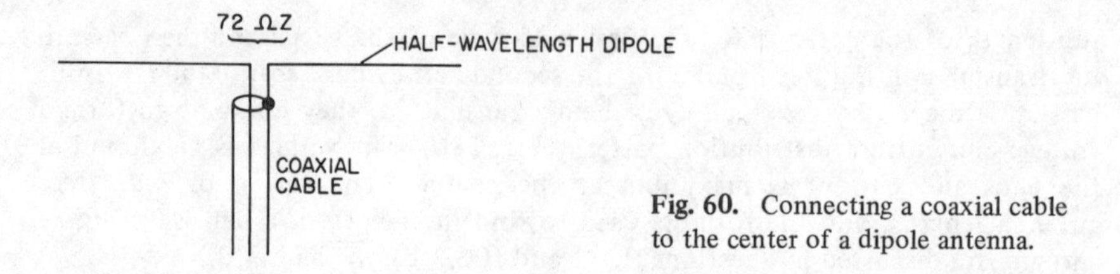

Fig. 60. Connecting a coaxial cable to the center of a dipole antenna.

straight radiation angle. However, if the antenna is placed close to the ground, the reflection from the ground causes the radiation pattern and the radiation angle to shift upwards (Fig. 59c). The value of impedance offered by the antenna is mainly that of the resistance offered to the radiation of signals (radiation resistance). At the center of a half-wave antenna in free space radiation resistance is approximately 72 Ω. The impedance (radiation resistance) offered by the dipole will be about 72 Ω if the half-wave antenna is split into two equal parts and fed at the center. Figure 60 illustrates a coaxial cable connected to the center feedpoint of a dipole antenna.

Figure 61 illustrates how the polarization of the radiating electric field from an antenna is determined by the position of the radiating element. Radiation is vertical for a vertical antenna and horizontal for a horizontal antenna.

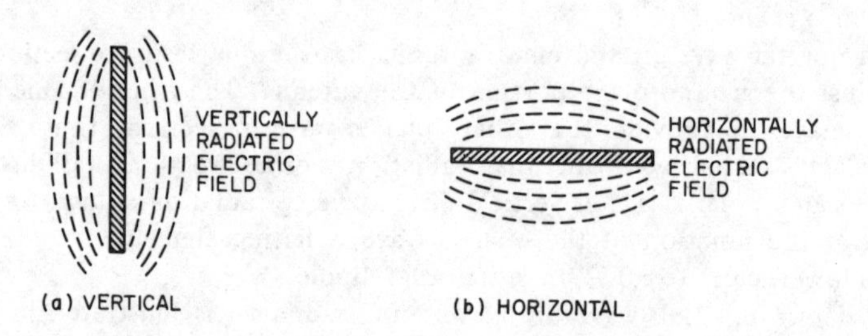

Fig. 61. Polarization of the radiated signal.

The construction of a quarter-wave ground plane antenna is shown in Fig. 62. The quarter-wave radiating element is vertically mounted. Four (or more) quarter-wave elements radiate out from the bottom in a horizontal direction to form an artificial ground (called a ground plane) against which the vertical radiating element operates.

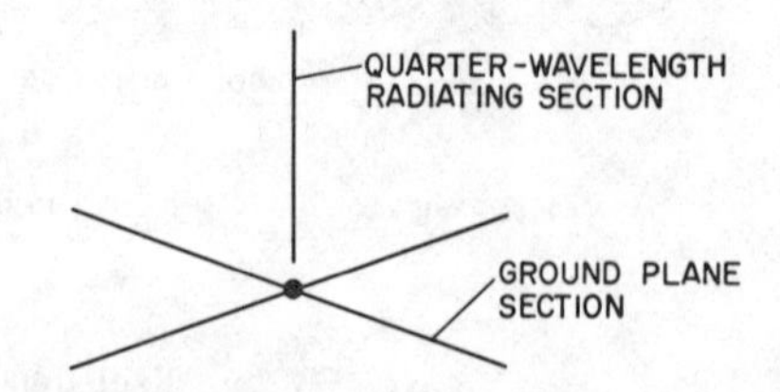

Fig. 62. Construction of a quarter-wave ground plane antenna.

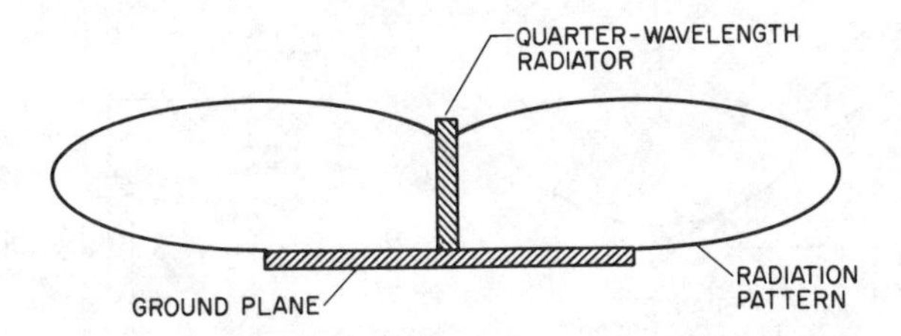

Fig. 63. Radiation pattern of a quarter-wave ground plane antenna.

A quarter-wave ground plane antenna, shown in Fig. 63, radiates equally in all directions to provide an omnidirectional pattern. Because the radiating element is vertical, the polarization is vertical. (See Fig. 61.) If a transmission line is connected to the quarter-wave ground plane antenna as shown in Fig. 64, the impedance of the antenna will be about 30 Ω.

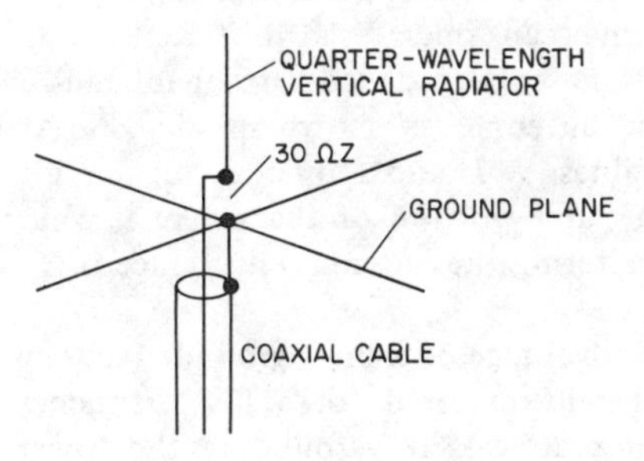

Fig. 64. Connecting a coaxial cable to a quarter-wave ground plane antenna.

H.3.3 How may a center-fed, horizontal, nonresonant antenna be used on several HF bands?

Answer: If the antenna is fed by a high-impedance open-wire or ribbon-type parallel-conductor transmission line, a transmatch should be used between the transmitter and the transmission line to match the output of the transmitter to the transmission line over the different HF bands. If the antena is fed by a low-impedance coaxial cable, traps should be used at specific distances in the antenna to maintain the antenna impedance at the different HF bands.

Discussion: Center-fed resonant half-wave horizontal antennas are illustrated in Figs. 58 and 60. An antenna is resonant when its length is either a half-wavelength or a multiple of a half-wavelength long at the frequency being applied. A nonresonant antenna is one whose length is not a half-wavelength or a multiple of a half-wavelength at the frequency being applied (Fig. 65).

When an antenna is used for several HF bands, its impedance changes with changes in the applied frequency. The fixed transmission line impedance will not match the changing antenna impedance at all bands. A high impedance parallel-conductor transmission line will minimize the effect of the changing antenna impedance. A transmatch is placed between the output of the transmitter and

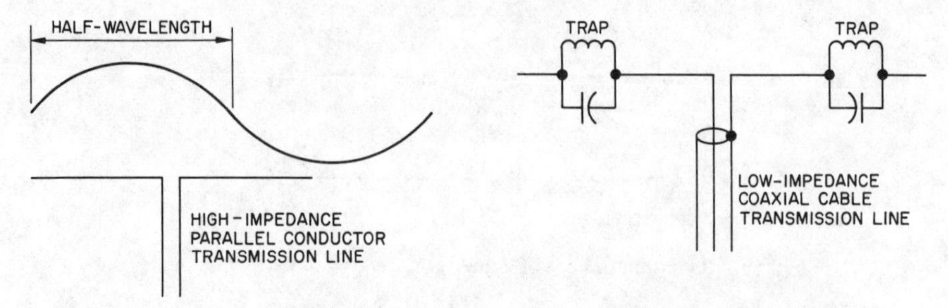

Fig. 65. Nonresonant horizontal dipole antenna center-fed with high-impedance parallel conductor transmission line.

Fig. 66. Nonresonant horizontal dipole antenna with traps, center-fed with low-impedance coaxial cable transmission line.

the parallel-conductor transmission line to tune the transmitter output system to resonance and minimize radiation of harmonic frequencies. (Refer to question I.2.7 in Chapter 2 for a discussion of a transmatch.)

A dipole antenna can operate at different bands of frequencies and maintain a constant value of impedance if resonant tank circuits are inserted at specific points in the antenna, as shown in Fig. 66. At specific frequencies, determined by the values of L and C in the trap circuit, the trap will act to isolate the remaining outer portion of the antenna, which effectively shortens the antenna without altering the antenna impedance offered at the transmission line feedpoint.

H.3.4(a) What is the advantage of using a grounded antenna?

Answer: The height required for a low frequency vertically polarized antenna would be impractical. By grounding the lower end of the vertical radiator we can use a quarter-wavelength in place of a half-wavelength since the ground acts as a reflector to provide the other half of the antenna.

H.3.4(b) What is the advantage of a multielement array antenna?

Answer: A multielement array provides increased gain and directivity.

Discussion: By placing additional elements in front of, or in back of a dipole, the signal radiating from the dipole can be reflected by these additional elements. The elements use the energy reflected from the dipole, called the driven element, thus they are called *parasitic* elements. As shown in Fig. 67(a), placing a slightly shorter parasitic element in front, and a slightly larger parasitic

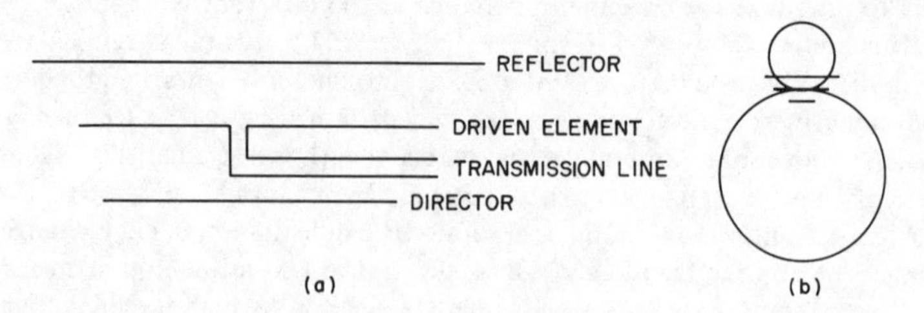

Fig. 67. (a) Arrangement of parasitic elements and (b) radiation pattern of a multielement array.

element in back of the driven element creates a directive radiation pattern, shown in Fig. 67(b). The action of the parasitic elements in reradiating the signal from the driven element can provide power gains of approximately four times (6 dB) that of the driven element alone. Using additional parasitic elements can provide further power gain.

H.3.5(a) What is standing wave ratio (SWR)?

Answer: When the transmission line and the antenna are not perfectly matched reflected standing waves will be present on the transmission line. The ratio between the maximum and minimum values of reflected or incident current (or voltage) on the transmission line is the standing wave ratio (SWR).

Discussion: When a mismatch exists between the impedance at the feedpoint of the antenna and that of the transmission line, voltage and current are reflected from the antenna feedpoint back to the transmitter along the transmission line. This reflection appears as stationary (standing) waves along the transmission line. (The power going out from the transmitter is called the *incident* power; the power coming back towards the transmitter is called the *reflected* power.) The ratio between the maximum and minimum values of reflected or incident current, or voltage, is the SWR and can be found by using either of the following formulas:

$$\text{SWR} = \frac{E_{max}}{E_{min}} \quad \text{or} \quad \text{SWR} = \frac{I_{max}}{I_{min}}$$

The values of incident and reflected power can be read by using a special bridge circuit called an SWR bridege or *reflectometer* placed between the transmitter and the transmission line.

H.3.5(b) How can the SWR on a transmission line be determined from the incident and reflected voltages?

Answer: SWR can be found with the following formula, using either the incident or the reflected voltage:

$$\text{SWR} = \frac{E_{max}}{E_{min}}$$

(See the discussion of H.3.5(a) above.)

H.3.6 What major characteristics determine the characteristic impedance of a parallel-conductor transmission line?

Answer: The characteristic impedance is determined by the distance between conductors and the diameter of the conductors.

Discussion: The characteristic impedance of a parallel conductor is determined primarily by the spacing between the conductors and the diameter of the wire used. Two No. 14-gauge wires separated by 2-inch spacers provides a typical impedance of about 500 Ω. Using 4-inch spacers increases the impedance to approximately 575 Ω. Two No. 16-gauge wires separated by 2-inch spacers provides an impedance of approximately 550 Ω. (Refer to question H.2.3(b) in Chapter 2 for more information on parallel-conductor transmission lines.)

Parallel-conductor line losses result mainly from resistance of the conductor and to a lesser degree from radiation from the line. These losses are usually given in attenuation in decibels per unit length. For 500-Ω open wire transmission line the loss is approximately 0.03 dB per 100 feet at 7 MHz. As the frequency increases the losses also increase, thus at 14 MHz the losses are approximately 0.05 dB per 100 feet.

H.3.7 What are the major factors that determine the characteristics of a coaxial transmission line?

Answer: The major factors include the distance between the inner and outer conductors, the diameter of the conductors, and the type of insulating material used between the conductors.

Discussion: The characteristic impedance of coaxial cable is determined by the combination of the size of the inner conductor, the thickness of the outer conductor of braided shielding, and the type of material used between the two conductors. Coaxial cables used for low power have a thin center conductor of approximately No. 20-gauge wire, an outer braid conductor approximately ¼ inch in diameter, and a coating of weatherproof plastic. The dielectric (insulating) material between conductors is usually polyethylene. For higher power, the center conductor may be three No. 18-gauge wires stranded together, and the outside braid conductor may be approximately 3/4 inch in diameter. A heavier center conductor and outer conductor with a wider spacing between conductors provides a coaxial cable with the same impedance as that of the smaller cable but capable of withstanding much higher voltages. (Refer to question H.2.3(b) in Chapter 2 for additional information on coaxial cable transmission lines.)

Coaxial cable line losses result mainly from resistance of the conductor, dielectric loss and radiation loss. Dielectric loss in a coaxial cable results from the loss in power required to charge the capacity between conductors, which is greater in coaxial cable than in parallel-conductor transmission line. The loss for RG 8/U coaxial cable is approximately 1 dB per 100 feet at 7 MHz and approximately 1.5 dB per 100 feet at 14 MHz.

I. Radio Communication Practices

I.3.1(a) What one instrument will give more information, more accurately, for the proper adjustment of a radiotelephony transmitter, than will any other collection of instruments generally available to amateur operators?

Answer: An oscilloscope.

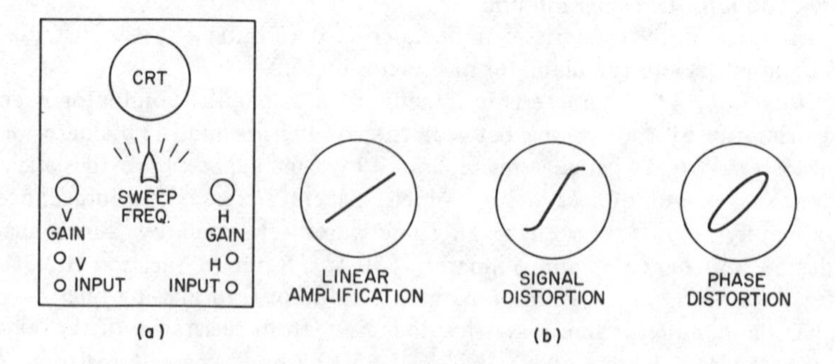

Fig. 68. (a) Basic oscilloscope; (b) oscilloscope test patterns for linearity testing.

Discussion: The oscilloscope is capable of displaying on the face of a cathode ray tube (CRT) a graphic display that is a reproduction of the waveform(s) available in the circuit to which the oscilloscope is attached. A basic oscilloscope is illustrated in Fig. 68(a). The CRT develops a beam of electrons that strike the front of the phosphor-coated tube to develop a dot of light. By controlling the beam in combined vertical and horizontal positioning, the beam can be made to "draw" a waveform on the face of the CRT.

Two internal amplifiers, vertical (V) and horizontal (H), are used to amplify externally applied signals to a level which will be sufficiently strong to control the beam in the CRT. When the signal input is to be compared to a value of time, such as in frequency observing, a built-in sweep circuit draws the beam horizontally across the face of the CRT in a fraction of time corresponding to the frequency under observation. For example, if one wishes to observe a cycle of ac line current, he would have the beam sweep across the face of the CRT in 1/60th of a second.

I.3.1(b) How is the oscilloscope used?

Answer: By properly connecting the oscilloscope to the circuits being observed or tested, the oscilloscope can display: amplifier linearity, percentage of modulation or overmodulation, frequency ratios, PEP test patterns using two-tone tests, and for over- or underdriven linear amplifiers using two-tone tests.

Discussion: To test an audio amplifier for a linear relationship between the input and output signals, a single audio tone sine wave signal is applied to the input of the amplifier. The vertical input terminals of the oscilloscope are connected to the input of the audio amplifier, and the horizontal input terminals of the oscilloscope are connected to the output of the audio amplifier. (See Fig. 68b). Assuming no distortion in the oscilloscope amplifiers and equal gain in both the vertical and horizontal amplifiers, the oscilloscope display will be a 45°-line if the amplifier under test is linear. If there is distortion in the amplifier, the line will be bent, usually at the ends. Phase distortion (phase shift) produces an elliptical pattern.

By applying the output rf signal of a transmitter to the vertical amplifier and using the internal oscilloscope sweep generator, the modulation waveform of the output of an AM transmitter can be observed. As shown in Fig. 69, at zero percent the carrier is completely unmodulated. With a single-tone modulating signal that is one-half the power of the carrier signal, the carrier fluctuates from

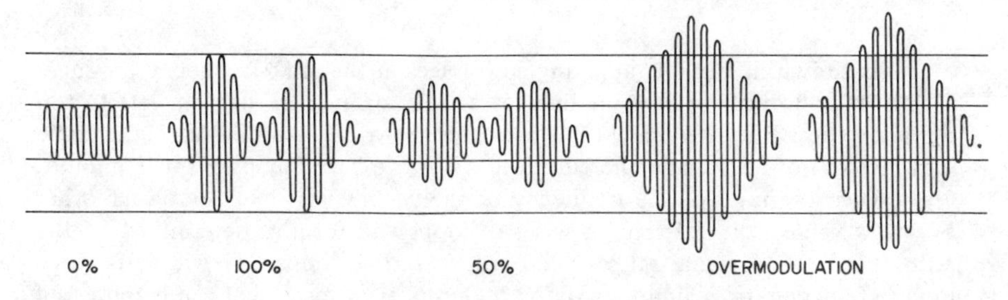

Fig. 69. Amplitude modulation waveforms.

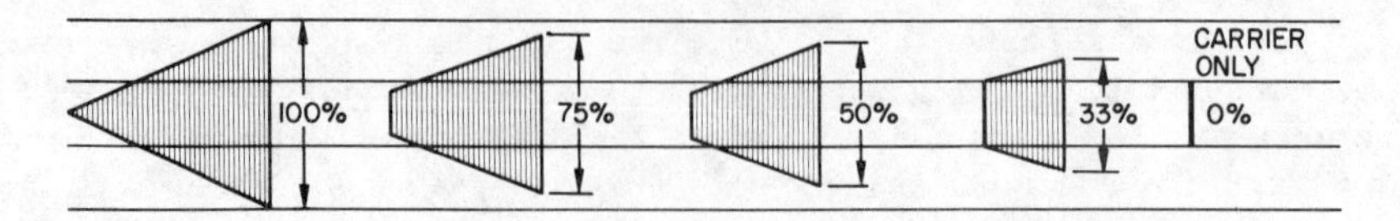

Fig. 70. Trapezoidal amplitude modulation waveforms.

zero to twice its value to provide 100% modulation. With lesser values of percentage modulation, the modulation waveform still provides an envelope but it is in proportion to the modulating power. That for 50% modulation is also shown in Fig. 69. When the modulating power exceeds one-half of the carrier power modulation exceeds 100% which overmodulates the carrier and creates the distorted output shown in Fig. 69.

In another method of comparing percentage of modulation, the oscilloscope is connected so as to obtain a trapezoidal pattern. The trapezoidal waveshape shown in Fig. 70 appears on the oscilloscope screen when the modulated rf carrier is applied to the vertical plates of the CRT and the audio modulating signal is applied to the horizontal amplifier input. At 100 percent modulation, a wedge-shape is produced. As the value of modulation is reduced, the point of the wedge is removed, and the amplitude of the pattern is proportionally reduced.

The frequency of a signal can be checked by using the oscilloscope to develop *Lissajous figures.* Lissajous figures are the waveforms that appear on the CRT when one ac voltage is applied to the vertical circuits and another ac voltage is applied to the horizontal circuits. The voltages combine to form figures on the screen, which indicate the relationship of the frequencies of the two ac voltages.

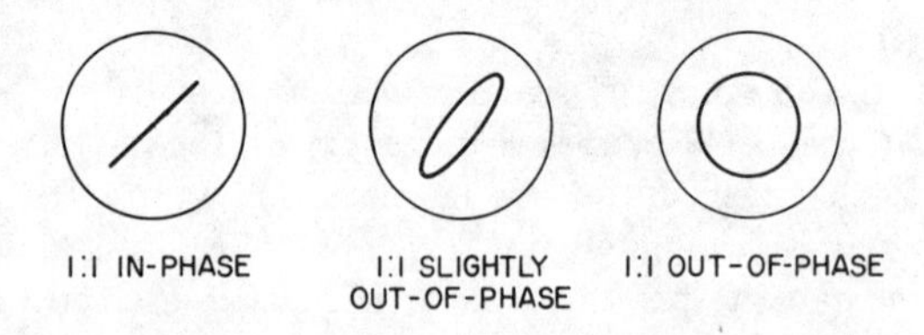

Fig. 71. Lissajous figures at a 1:1 frequency ratio.

As shown in Fig. 71, applying ac voltages of the same frequency produces a circular, elliptical, or straight line (at a 45° angle, from bottom left to top right) on the CRT. The straight line indicates the voltages are in-phase; the circle indicates out-of-phase, and the elliptical is out-of-phase but by a small amount.

When the ratio of the frequency of the two signals are close to each other by a low value, 2:1, 3:1, etc., a series of loops will form as determined by the ratio (Fig. 72). A signal ratio of 2:1 will form two loops, 3:1 will form three loops, etc. The ratio is limited only by the number of loops that can be observed on the screen; it may go as high as 20:1 or more. In addition, if there is a frequency or phase difference, the loops will tend to roll making it difficult to

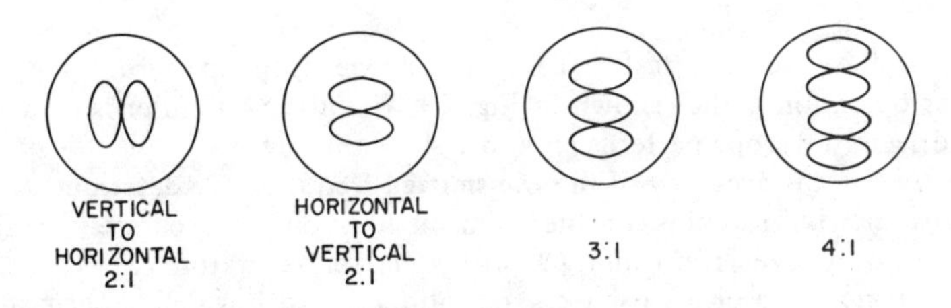

Fig. 72. Lissajous figures at different frequency ratios.

count the ratio. The ratio can be determined as being either vertical-to-horizontal or vice versa by the position of the loops, as shown in Fig. 72.

The ratio of peak envelope power (PEP) to the average power in an SSB signal depends primarily on the waveshape of the audio modulating signal and the intermodulation products. The strength of the modulating audio signal directly affects the ratio of PEP to average power. Estimating the characteristic of different human voices is difficult; the usual ratio is estimated to range between 1.5:1 to 3:1. For testing purposes, two rf signals are placed 1000 Hz apart through the rf amplifier system. The resultant output pattern shown in Fig. 73, provides a PEP that is two times the average power, a 2:1 ratio.

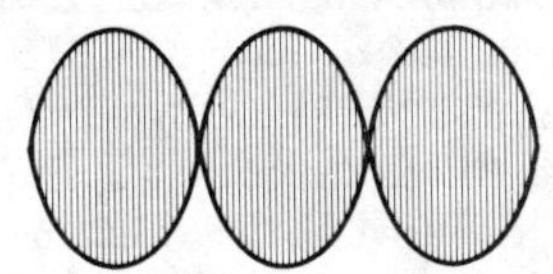

Fig. 73. Two-tone test pattern of a linear amplifier.

Application of a two-tone test signal can be used to check an SSB linear amplifier for proper operation, assuming all operating voltages are correct and input and output impedances are properly matched. To obtain the two-tone test signal, a 1000-Hz audio generator output signal is applied to the microphone input and the carrier insertion level is turned up until, with both controls (the output of the audio generator and the carrier insertion level) properly adjusted, the oscilloscope signal appears, as shown in Fig. 73. The rf from the output is picked up through a two- or three-turn link and coupled to the vertical input of the oscilloscope. The internal sweep is used and synchronized at 60 Hz.

The amplifier grid bias is adjusted for the best possible crossover setting, where the two signals meet at the zero crossover line. Adjusting grid drive, grid bias, and loading of the output tank circuit will provide the proper shape of the test pattern, which appears as shown in Fig. 73.

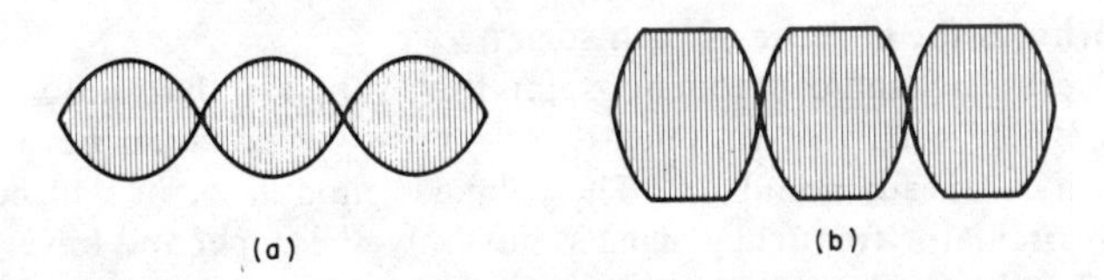

Fig. 74. Two-tone test patterns (a) underdriven and (b) overdriven linear amplifiers. linear amplifiers.

With the two-tone test signal applied, the scope pattern for optimum amplifier operation is that shown in Fig. 73. When the amplifier is underdriven or overdriven, the scope patterns are those shown in Fig. 74.

I.3.2 How can the accuracy of the transmitted frequency be determined?

Answer: By listening to the transmissions on an accurately calibrated receiver, or using a receiver calibrated with a marker generator.

Discussion: Receivers using crystal calibrators with which to accurately set the dial can be used to accurately determine the transmitted frequency. For receivers not having a built-in crystal calibrator an external crystal calibrator (also referred to as a marker generator) can be used to apply the crystal-controlled frequencies to the receiver for dial calibration.

I.3.3(a) How is the input power to the plate circuit of the final amplifying stage of an SSB transmitter determined?

Answer: The metered values of plate voltage is multiplied by the highest reading metered value of plate current of the final amplifying stage while a modulated signal is being emitted. (See the discussion of D.3.8(c) above.)

I.3.3(b) If power to the antenna is also supplied by the driver stage, how is the power determined?

Answer: All power supplied from any stage must be included as part of the input power rating. The power of a driver stage is added to that of the final amplifying stage to determine the total power input.

Discussion: In most SSB transmitters, the final amplifying stage has less power output than the legal limit. It is common practice to have the SSB transmitter drive a high-power linear amplifier to achieve full legal power output. When the final amplifying stage is a separate linear amplifier, the output of the transmitter is the driver power. With this arrangement, the driver power must be added to the linear amplifier power to derive the total input power supplied to the final amplifier.

I.3.4(a) What are the basic stages of a filter system SSB transmitter?

Answer: A block diagram illustrating the basic stages of a filter system SSB transmitter is shown in Fig. 75.

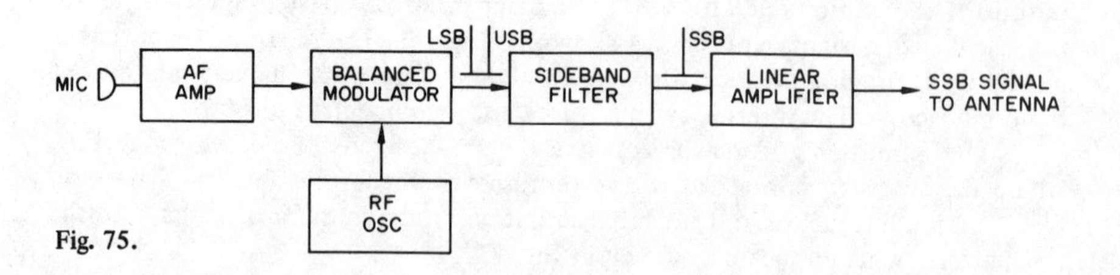

Fig. 75.

I.3.4(b) What is the purpose of each stage?

Answer: The audio frequency amplifier supplies the audio modulating signal and the rf oscillator supplies the rf carrier signal; then both signals are applied to a balanced modulator. The balanced modulator output contains no audio or rf oscillator frequency signals, but only the upper and lower sidebands developed by the mixing of the two signals. The two sidebands are applied to a specially designed filter circuit that passes only the desired band of frequencies.

The output of the filter, either upper or lower sideband, is the SSB signal. The SSB signal is then applied to a linear power amplifier stage whose output is applied to the antenna.

I.3.4(c) What are the basic stages of a phasing system SSB transmitter?

Answer: A block diagram illustrating the basic stages of a phasing system SSB transmitter is shown in Fig. 76.

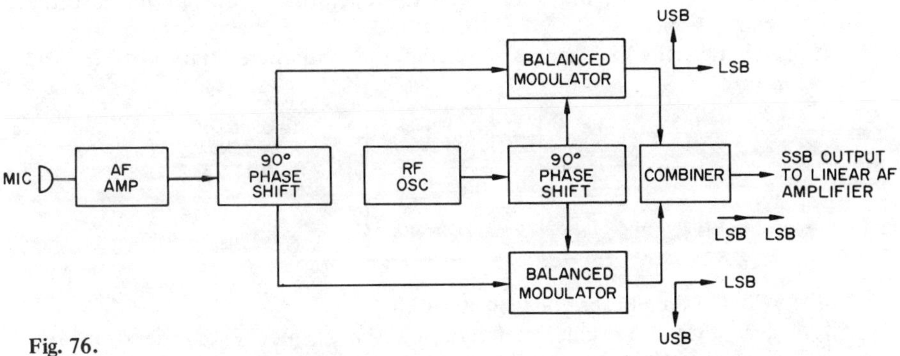

Fig. 76.

I.3.4(d) What is the purpose of each stage?

Answer: The audio frequency amplifier supplies the modulating audio frequency which is applied to a phase shifting network with equal value but 90° out-of-phase signals applied to balanced modulators. The output of the rf oscillator is also applied through 90° phase shift networks to both balanced modulators. The output of each balanced modulator will be an upper and lower sideband 90° out-of-phase, with each modulator output having the upper sidebands 180° out-of-phase and the lower sidebands in-phase. The 180° out-of-phase signals cancel each other, whereas the two lower sideband signals reinforce each other to provide a single sideband output.

Discussion: If USB signals are desired in the output, either the out-of-phase audio signals to the balanced modulator are reversed or the out-of-phase rf signal is reversed.

I.3.5(a) What are the basic stages of a transmitter for F3 emission?

Answer: A block diagram illustrating the basic stages for an indirect FM (F3) transmitter is shown in Fig. 77.

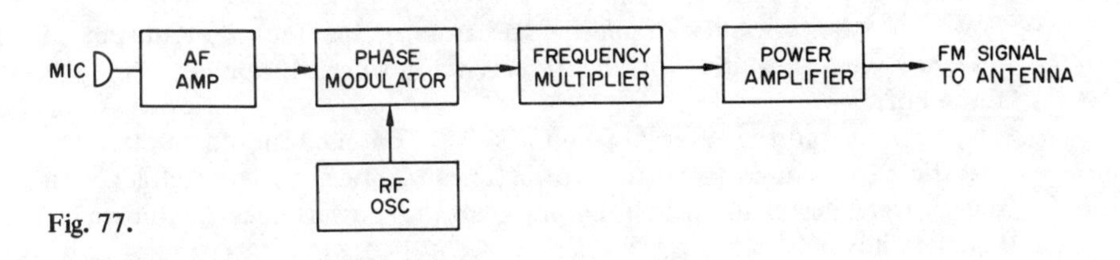

Fig. 77.

I.3.5(b) What is the purpose of each stage?

Answer: The audio frequency amplifier supplies the audio modulating signal and the rf oscillator supplies the rf signal, both of which are applied to the phase modulator. The phase modulator causes the audio modulating signal to vary the phase, which in turn varies the frequency, of the rf oscillator signal. The FM signal output frequency of the phase modulator is then changed by the frequency multiplier to a higher value frequency that falls in the amateur operating frequency bands. The final high frequency signal is then amplified by a power amplifier and applied to an antenna. (It should be noted that the frequency multiplier not only multiplies the operating frequency but also the frequency deviation.)

I.3.6(a) What are the basic stages of an amplitude modulated transmitter?

Answer:

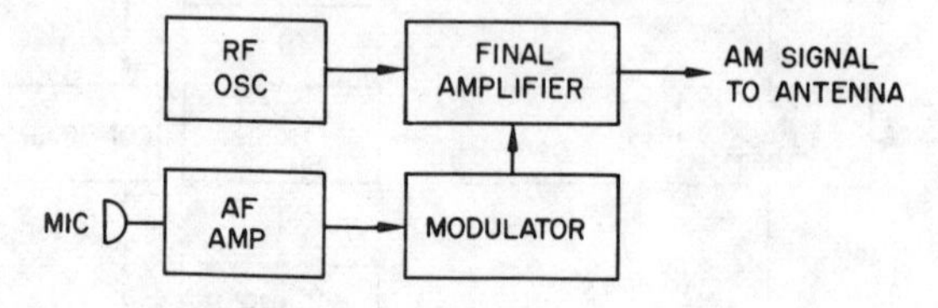

I.3.6(b) What is the purpose of each stage?

Answer: The rf oscillator determines the frequency of operation and provides an unmodulated carrier to the final amplifier. The audio frequency amplifier output is applied to the modulator. The modulator causes the unmodulated carrier and audio frequency to combine and have the carrier frequency amplitude modulated both positively and negatively to provide an amplitude modulated output signal.

I.3.7 What precautions should be taken in the construction and operation of amateur radio equipment to avoid the danger of electrical shock?

Answer: The major precautions are:

1. The equipment should be enclosed to prevent accidental contact with exposed high voltage points.
2. All metal chassis should be connected directly to ground, preferably to the water pipe (metal, not plastic) at the inlet to the building. If this is not practical, the chassis should be returned to electrical ground through an approved grounded conductor as noted in the following precaution.
3. The wall outlet used to supply ac voltage should be an Underwriters Laboratories approved 3-prong grounded outlet; and the power cable from the equipment to the wall outlet should be a matching 2-wire plus groundwire (3-conductor) cable.
4. Power supplies should use *bleeder* resistors across the filter output to discharge the capacitors of the filter circuit when the power supply is turned off.
5. The antenna and the external feed lines to the antenna should be placed at a sufficient distance from any power lines to prevent their contact with the power lines should the antenna or the power lines be blown or knocked down.

I.3.8 Define receiver sensitivity and selectivity.

Answer: Sensitivity is the ability of the receiver to respond to weak signals. Selectivity is the ability of the receiver to select a particular signal while rejecting all others.

I.3.9(a) In repeater operation, what is the effect of desensitization?

Answer: Desensitization is the reduction of the sensitivity of the repeater receiver.

I.3.9(b) What is the major cause of desensitization?

Answer: The major cause of desensitization is the repeater receiver being overloaded by the signal from the repeater transmitter.

I.3.10(a) What is receiver cross modulation?

Answer: An undesired strong signal close to the frequency to which the receiver is tuned will be amplified by the rf amplifier stages to create a signal whose strength is such that it will modulate the desired signal.

I.3.10(b) What is transmitter intermodulation?

Answer: Transmitter intermodulation is when two transmitters close to each other may have the fundamental or harmonic frequency of one transmitter modulate the signal being generated in the other transmitter to provide *intermodulation.*

I.3.11 What are the major sources of receiver noise?

Answer: The major noise sources are thermal noise, shot noise, man-made noise, and atmospheric noise.

Discussion: Random movements of electrons in the circuit wiring create minute voltages that cause background noise, called thermal noise. Vacuum tubes and transistors develop *shot effect* noise due to the random movement of current carriers and recombination of majority and minority current carriers in diodes and transistors. Arc-discharges such as those due to spark-plug gaps of automotive ignition systems and at the brushes of small universal motors used in appliances provide man-made noise. Discharges due to lightning are considered atmospheric noise.

I.3.12(a) What is the basic method of detecting AM signals?

Answer: A diode detector is used, with the rectified output filtered to provide an audio signal.

I.3.12(b) What is the basic method of detecting CW signals?

Answer: By mixing (beating, or heterodyning) a special oscillator, called a beat frequency oscillator (BFO), with a diode detector. When the incoming signal beats with the BFO output, the product is detected by the diode to provide an audible CW signal.

I.3.12(c) What is the basic method of detecting SSB signals?

Answer: As with the CW detector, a special oscillator beats with the incoming signal at the diode detector, and the product is an audible signal. A SSB detector is most often referred to as a product detector.

I.3.12(d) What is the basic method of detecting FM signals?

Answer: A frequency discriminator circuit is used in which there will be no output when the FM signal is at the center or resting frequency. When the signal varies above and below the center frequency, the output of the discriminator provides positive and negative voltages at the frequency with which the signal varies. The amplitude of the output voltage is directly proportional to the amount of frequency deviation.

A newer type FM detector utilizes a phase lock loop (PLL) in an integrated circuit. The PLL basically detects by using a local oscillator whose phase (frequency) is varied by the incoming FM signal. The phase or frequency variations produced by the FM signal are detected to provide an audio output.

Chapter 4

STATION OPERATION

If the student has studied the text in this book diligently and practiced the code, he probably will have passed the FCC tests and his license is either on its way in the mail or in his hands. Once the reader has his license, how does he best use it? The types of equipment he can use are best discussed between him and his amateur radio acquaintances. He can learn from their experience, since it is a combination of experience, his desires, and his ability to pay that determines what his station equipment will be. Whether he builds his own, buys second-hand, or goes for broke, only he can determine what is best for him.

What this text can tell him is how best to start to develop the proper procedure for operating his station. Each operator develops his own approach, and if the reader stays with CW, he will develop his own "fist," or style of transmitting CW. Whether he uses phone or CW, there is a standard approach to operating the station. For operating teletype (TTY), FM repeaters, television, or any other specialized form of transmission, he should read the special literature available on these subjects and, more important, observe other amateurs while they are operating these specialized forms of amateur radio transmission.

There are many different aspects of operating available for amateurs in addition to the different forms provided by the types of transmission chosen, that is, CW, phone, TTY, television, etc. He can also choose from among any of the following types of activities: random calling for anyone on the band who cares to work him (calling CQ to "ragchew"); working acquaintances on a "schedule"; working a group of amateurs on a scheduled "roundtable"; working stations in foreign countries (working "DX"); working with a group of amateurs on a network scheduled to handle traffic (noncommercial messages for no fee); working with local officials for Civil Defense operations; operating to qualify for special awards offered by various magazines, societies, and local groups, such as WAS (worked all states); working amateur contests that take place at regular intervals, the best known being Field Day, where individuals or groups can compete for awards; operating with amateur satellites in orbit; working with the army as part of the Military Affiliate Reserve System (MARS); and participating in the American Radio Relay League (ARRL) activities. Despite this long list, there are still more forms of operating. An example of a specialized operation is the establishment of an amateur radio station at a remote area anywhere on the globe so that DX operators can "log" the call sign assigned to the remote area.

Operating Aids

As with any other endeavor, short cuts are developed to make operating simpler, faster, and more reliable. Since CW was the first type of amateur operation and has the longest history, the majority of operating aids were created in use in CW. However, because CW aids say so much in so few words or letters and because of their familiarity, many of the CW operating aids are used with phone transmission.

Probably the most frequently used form of aid is the Q signal. This, too, is a form of code using three letters, the first one always being Q. When the Q signal is followed by a question mark, it signifies an inquiry; when the question mark is omitted, it signifies an answer or a statement. The list of Q signals appears in Table 6.

Table 6. Q Signal Abbreviations and Their Use.

Signal	Question	Answer
QAV	Are you calling me?	I am calling . . .
QRG	Will you tell me my exact frequency (or that of . . .)?	Your exact frequency (or that of . . .) is . . . kHz (or MHz).
QRH	Does my frequency vary?	Your frequency varies.
QRI	How is the tone of my transmission?	The tone of your transmission is . . . (1. Good 2. Variable 3. Bad).
QRK	What is the readability of my signal (1 to 5)?	The readability of your signal is . . . (1 to 5).
QRL	Are you busy?	I am busy. Or (I am busy with . . .). Please do not interfere.
QRM	Are you being interfered with?	I am being interfered with.
QRN	Are you troubled by atmospherics?	I am troubled by atmospherics.
QRO	Shall I increase power?	Increase power.
QRP	Shall I decrease power?	Decrease power.
QRQ	Shall I send faster?	Send faster (. . . words per minute).
QRRR	American Radio Relay League (abbreviation for land SOS).	This is a distress call for emergencies only.
QRS	Shall I send more slowly?	Send more slowly (. . . words per minute).

(*continued*)

Table 6. Q Signal Abbreviations and Their Use (*continued*).

Signal	Question	Answer
QRT	Shall I stop sending?	Stop sending.
QRU	Have you anything for me?	I have nothing for you.
QRV	Are you ready?	I am ready.
QRW	Shall I inform . . . that you are calling him on . . . kHz (or MHz)?	Please inform . . . that I am calling him on . . . kHz (or MHz).
QRX	When will you call again?	I will call you again at . . . hours (on . . . kHz or MHz).
QRY	What is my turn?	Your turn is number . . . .
QRZ	Who is calling me?	You are being called by . . . (on . . . kHz or MHz).
QSA	What is the strength of my signals (1 to 9)?	The strength of your signal is . . . (1 to 9).
QSB	Are my signals fading?	Your signals are fading.
QSD	Is my keying defective?	Your keying is defective.
QSG	Shall I send . . . messages at a time.	Send . . . messages at a time.
QSK	Can you hear me between your signals and if so can I break in on your transmission?	I can hear you between my signals; break in on my transmission.
QSL	Can you acknowledge receipt?	I am acknowledging receipt.
QSM	Shall I repeat the last message I sent you, or some previous message?	Repeat the last message you sent me (or message(s) number(s) . . .).
QSN	Did you hear me (or . . .) on . . . kHz (or MHz)?	I heard you (or . . .) on . . . kHz (or MHz).
QSO	Can you communicate with . . . direct or by relay?	I can communicate with . . . direct (or by relay through . . .).
QSP	Will you relay to . . . free of charge?	I will relay to . . . free of charge.
QST	———	A general call preceding a message addressed to all hams.
QSU	Shall I send or reply on this frequency (or on . . . kHz or MHz)?	Send or reply on this frequency (or on . . . kHz or MHz).

* From *Building the Amateur Radio Station,* J. Berens and J. Berens. Rochelle Park, N.J.: Hayden Book Co., 1965.

Table 6. Q Signal Abbreviations and Their Use (*continued*).

Signal	Question	Answer
QSV	Shall I send a series of *V*s on this frequency (or . . . kHz or MHz)?	Send a series of *V*s on this frequency (or . . . kHz or MHz).
QSW	Will you send on this frequency (or on . . . kHz or MHz)?	I am going to send on this frequency (or . . . kHz or MHz).
QSX	Will you listen to . . . on . . . kHz (or MHz)?	I am listening to . . . on . . . kHz (or MHz)
QSY	Shall I change to transmission on another frequency?	Change to transmission on another frequency (or . . . kHz or MHz).
QSZ	Shall I send each word or group more than once?	Send each word or group twice (or . . . times).
QTA	Shall I cancel telegram No. . . . as if it had not been sent?	Cancel telegram No. . . . as if it had not been sent.
QTB	Do you agree with my counting of words?	I do not agree with your counting of words; I will repeat the first letter or digit of each word or group.
QTC	How many telegrams have you to send?	I have . . . telegrams for you or for . . .
QTH	What is your location?	My location is . . .
QTR	What is the correct time?	The correct time is . . .

*From *Building the Amateur Radio Station,* J. Berens and J. Berens. Rochelle Park, N.J.: Hayden Book Co., 1965.

One of the first things a newly operating amateur wants to know is how well he is being received. He also knows that the other operator has the same question to ask of him. Because this is such a common question, a standard reporting system, now in universal use, was worked out. Called the RST system [previously discussed in Chapter 2, question C.2.2(a)] it provides three areas of signal identification: *R* for reaeability, *S* for signal strength, and *T* for tone. Although RST provides for three reports, the tone report is usually given only when the signal received is poor or defective in some manner. For normal operation, the RST report will usually provide only the *R* and *S* values. For example, a perfectly readable, moderately strong signal of purest dc tone will provide a transmission of "RST 57" with the report on phone being "I read you 5 by 7."

When operating CW, it becomes tiresome to sepll out all words, particularly those used many times. As a result many words have been abbreviated to form a sort of CW "shorthand," shown in Table 7. In addition to these standard

Table 7. Common CW Abbreviations and Their Meanings.

AA	All after
AB	All before
ABT	About
ADR	Address
AGN	Again
ANT	Antenna
BCI	Broadcast interference
BCL	Broadcast listener
BK	Break; break me; break in
BN	All between; been
BUG	Semi-automatic key
C	Yes; correct
CFM	Confirm; I confirm
CK	Check
CL	I am closing my station; call
CLD-CLG	Called; calling
CQ	Calling any station
CUD	Could
CUL	See you later
CUM	Come
CW	Continuous wave (i.e., radio telegraph)
DE	From, send by
DLD-DLVD	Delivered
DX	Distance, foreign countries
ES	And, &
FB	Fine business; excellent
G	Repeat
GA	Go ahead (or resume sending)
GB	Good-by
GBA	Give better address
GE	Good evening
GG	Going
GM	Good morning
GN	Good night
GND	Ground
GUD	Good
HI	The telegraphic laugh; hi
HR	Here; hear
HV	Have
HW	How
LID	A poor operator
MA, MILS	Milliamperes
MN	Minute
MSG	Message; prefix to radiogram
N	No
NCS	Net control station
ND	Nothing doing
NIL	Nothing; I have nothing for you
NM	No more
NR	Number, near
NW	Now; I resume transmission
OB	Old boy
OM	Old man
OP-OPR	Operator
OT	Old timer; old top
PBL	Preamble
PSE	Please
PWR	Power
PX	Press
R	Received as transmitted
RCD	Received
RCVR (RX)	Receiver
REF	Refer to; referring to; reference
RFI	Radio frequency interference
RI	Radio inspector
RIG	Station equipment
RPT	Repeat; I repeat
RTTY	Radioteletype
SASE	Self-addressed, stamped envelope
SED	Said
SIG	Signature; signal
SINE	Operator's personal initials or nickname
SKED	Schedule
SRI	Sorry
SVC	Service; prefix to service message
TFC	Traffic
TMW	Tomorrow
TNX-TKS	Thanks
TT	That
TU	Thank you
TVI	Television interference
TXT	Text
UR-URS	Your; you're; yours
VFO	Variable-frequency oscillator
VY	Very
WA	Word after
WAT	What
WB	Word before
WD-WDS	Word; words
WKD-WKG	Worked; working
WL	Well; will
WUD	Would
WX	Weather
XMTR (TX)	Transmitter
XTAL	Crystal
XYL (YF)	Wife (or married woman)
YL	Young lady
73	Best regards
88	Love and kisses

forms of abbreviation, CW operators will use individual common-sense abbreviations during a conversation. For example, UR SIGS QSB C U NEXT SKED means "Your signals are fading; we'll get together again on the next scheduled transmission."

Although not an aid as such, a well-known form of abbreviation is CQ, which translates "seek you" and means calling any station that wishes to reply.

In phone operation, phonetics poses a problem since many different letters sound alike. For example, B,C,D; pronounced phonetically is "bee, cee, dee." With a call sign such as W-BCD, the query is most likely to be, "Was that BCD or BCP?" To overcome such confusions, a standard international phonetic alphabet was derived (Table 8). With this alphabet, the call would be given as "W-BCD, Bravo, Charlie, Delta."

In addition to being a requirement, the station log can also be considered an operating aid. If filled out completely, it can be used as a reference for operators' names and for operating frequencies. Operators' call signs noted in the log can be used to find their local addresses for mailing purposes by refering to the two available directories listing amateur operators' addresses by their call sign. One directory is for United States amateurs only; the other includes all

Table 8. Phonetic Alphabet.

Letter	*Identifying word*	*Spoken as**
A	Alfa	AL fah
B	Bravo	BRAH voh
C	Charlie	CHAR lee (or SHAR lee)
D	Delta	DELL tah
E	Echo	ECK oh
F	Foxtrot	FOKS trot
G	Golf	golf
H	Hotel	HOH tell
I	India	IN dee ah
J	Juliett	JEW lee ETT
K	Kilo	KEY loh
L	Lima	LEE mah
M	Mike	mike
N	November	No VEM ber
O	Oscar	OSS cah
P	Papa	Pah PAH
Q	Quebec	Keh BECK
R	Romeo	ROW me oh
S	Sierra	See AIRRAH
T	Tango	TANG go
U	Uniform	YOU nee form (or OO nee form)
V	Victor	VIK tah
W	Whiskey	WISS key
X	X-ray	ECKS ray
Y	Yankee	YANG key
Z	Zulu	ZOO loo

* The symbols to be emphasized are in capital letters.

amateurs throughout the world. In filling out the log, it might be helpful in addition to, or in place of, logging local time to log or operate on Greenwich Mean Time (GMT). Greenwich Mean Time uses the time at Greenwich, England, as standard all over the world; thus all clocks set to GMT will read the same time at any place in the world. GMT logging is particularly helpful in operating DX.

CW Operation

One of the more common pitfalls in CW operation is the desire to speed up. The beginning operator, now that he has his license, feels compelled to catch up with the more experienced operators. In doing this the new operator invariably runs his letters together so that it is difficult to separate the letters, let alone the words, of the message being sent. His awakening to the run-ins may be the result of numerous QRS signals or perhaps of listening to a tape of his transmissions. The beginning operator should concentrate on a "clean fist" approach, making sure that he is not asked to repeat, that he does not have to stop due to an error, send a string of "E's," and repeat the message. In time he will acquire speed and accuracy and become a good operator. After acquiring speed using the conventional telegraph key, the operator, now experienced, can graduate to using "bugs," electronic keys, or any other device that helps to speed up his transmission. Although some operators are proud of their special style of transmitting, called a *swing*, most operators are proud that listeners cannot differentiate between their fist and transmission from a code machine.

Before transmitting, the operator should be sure that he will not interfere with other operators already using that frequency or operating close to that frequency. He should listen for several moments. Sometimes, due to the frequency being used, he will not be able to hear operator A but can hear the signal when operator B answers. Thus, the frequency originally thought clear is not really clear. Transmitting on that frequency would create interference.

When calling either to a specific station or for any station, the usual procedure is to call three times and give the call three times. W1AAU W1AAU W1AAU DE W1YCV W1YCV W1YCV or CQ CQ CQ DE W1YCV W1YCV W1YCV. To work a station in a particular area, the call would be CQ W1 CQ W1 CQ W1 DE W1YCV W1YCV W1YCV. To work a station in a particular city, the call would be CQ BOSTON CQ BOSTON CQ BOSTON, etc.

When answering a call the first time, the same procedure is used: the call letters are given three times, followed by K, go ahead. For example, W1YCV W1YCV W1YCV DE W1AAU W1AAU W1AAU K. The long return call provides an opportunity to tune on the station returning the call. Once contact is established, each call need be given only once: W1AAU DE W1YCV K.

Break-in procedures exist for joining a QSO already in progress. The most common procedure is to wait for a lull in the transmission and then transmit either $\overline{\text{BK}}$ or $\overline{\text{BT}}$, indicating that a "breaker" desires to join the QSO. If heard by either party in the QSO, he will stop at a convenient point and invite the breaker to join the QSO. Another form of break-in is used for a traffic handling net operation during transmission of a message. When the party copying the message misses a portion, he will tap the key a few times, transmitting a few dashes, which will be heard by the transmitting station between characters. This signal indicates that the break is due either to an error or to some other reason the breaker gives.

A period, dit dah dit dah dit dah, is cumbersome and so is not often used. Instead, the break sign, $\overline{BT}$, or the letter X is often used at the end of a sentence. The period used in decimals or those used for colon frequently are indicated by the letter R. For example, 2.5 feet is sent as 2R5 feet; 3:10 is sent as 3R10.

Phone Operation

In phone operation, a new operator's pitfall is again speed—being in a hurry to give the call without any phonetic pronunciation. It is necessary to speak clearly and directly into the microphone. To talk quickly, as one would to family or friends familiar with one's speech patterns, may be a problem on the air when talking to a stranger.

The operating courtesy for CW operation is used for phone operation, too. Before transmitting, one should listen carefully to be sure he will not interfere.

For phone operation, the same call procedure is used: call three times and identify three times. For example, W1AAU W1AAU W1AAU THIS IS W1YCV W1YCV W1YCV OVER. On a CQ call, giving the last of the three identifying calls phonetically will make it easier for a station to identify correctly. For example, CQ CQ CQ THIS IS W1YCV W1YCV W1 YANKEE CHARLIE VICTOR. When calling geographical area, operation is identical with that given for CW except that the location is pronounced instead of being given as initials or abbreviations, which are simpler for CW.

To answer a call, the same technique is used as that used for CW. For the initial contact the call signs are given three times; after contact, one time suffices.

In phone operation, one tends to use CW signals since they are abbreviated. Some abbreviations have come into common use, such as "holding a QSO"; however, it is best to stick to speech for routine operation. When using phone, CW abbreviations, such as "Roger—received that fine OM," should not be used.

Since phone operation tends to take longer than CW, the operator must watch that conversation doesn't exceed the 10-minute legal time limit for identifying the station. One must also take care not to operate back and forth without giving call signs. The rules call for identification at the beginning and end of each single transmission or exchange of transmissions not to exceed 10 minutes in duration. This last rule holds particularly true for stations using voice-operated switched (VOX) transmitters.

Break-in procedures for phone are similar to those used for CW; the only difference is that the word *break* is used.

Appendix I

FCC-TYPE EXAMS AND ANSWERS

FCC-Type Examination for a Novice Class License

1. In the FCC rules, the part governing amateur radio service is:
(a) Part 97. (b) Part 79. (c) Part 67. (d) Part 76. (e) Part 27.
2. Upon conviction, the maximum penalty for each day in which a violation of the FCC rules occurs can be for no more than:
(a) $50. (b) $100. (c) $500. (d) $1000. (e) $5000.
3. The FCC rules encourage the following two phases of the radio art:
(a) Communication and experimentation. (b) Communication and emergency traffic. (c) Experimentation and emergency traffic. (d) Emergency traffic and technical. (e) Communication and technical.
4. An amateur radio operator is defined as a person who:
(a) Holds an FCC amateur license. (b) Operates a licensed amateur radio station. (c) Operates an amateur radio station with no pecuniary interest. (d) Holds an FCC amateur license and operates an amateur station with no pecuniary interest. (e) Is an FCC licensed amateur interested in radio techniques.
5. After restructuring, a Communicator class license will be valid for:
(a) 1 year. (b) 5 years. (c) 3 years. (d) 2 years. (e) Indefinitely.
6. After restructuring, a Novice license will be:
(a) Not renewable. (b) Renewable 6 months after expiration. (c) Renewable 1 year after expiration. (d) Renewable upon expiration. (e) Issued for the life of the operator.
7. An amateur radio station license must be:
(a) Available on demand. (b) Posted anywhere within the room where the transmitter is located. (c) Posted in a conspicuous place in the room where the transmitter is located. (d) Kept on the person of the operator. (e) Worn on a conspicuous place of the amateur who is operating the station.
8. The person responsible for the proper operation of an amateur station is:
(a) The station licensee. (b) The operator licensee. (c) The control operator. (d) The designated station operator. (e) The station owner.
9. The definition of a control operator is:
(a) Anyone designated by the station licensee to operate the station. (b) Any

amateur radio operator designated by the station licensee to operate the station. (c) Whoever is operating the station. (d) Whoever is controlling the transmitter. (e) Anyone designated by an amateur operator to operate the station.

10. In filling out the log, standard information such as type of emission, location of station, input power:
 (a) Must be entered for each transmission. (b) Must be entered for each communication. (c) Can be entered once for each page of the log. (d) Can be entered for each day of operation. (e) Can be entered once at the beginning of the log book.
11. The log of a station must be preserved for a period of:
 (a) No less than 1 year following the last date of entry. (b) No more than 1 year following the last date of entry. (c) No less than 2 years following the last date of entry. (d) No more than 2 years following the last date of entry. (e) For as long as the station is in operation.
12. The holder of a Novice class license can operate an amateur radio station on:
 (a) 3700–3750 and 7100–7150 kHz; 21.0–21.5 and 28.5–29.0 MHz.
 (b) 3750–3800 and 7150–7200 kHz; 21.0–21.5 and 28.5–29.0 MHz.
 (c) 3700–3750 and 7100–7150 kHz; 21.1–21.2 and 28.1–28.2 MHz.
 (d) 3750–3800 and 7150–7200 kHz; 21.1–21.2 and 28.1–28.2 MHz.
 (e) 3750–3800 and 7100–7150 kHz; 21.1–21.2 and 28.1–28.2 MHz.
13. After restructuring, emission privileges authorized to Communicator class licensees will be:
 (a) Type A1 on all bands above 144 MHz. (b) Type F3 on all bands above 144 MHz. (c) Type A1 on all bands below 29 MHz. (d) Type F3 on all bands above 29 MHz. (e) Type F3 on all bands.
14. The proposed new Communicator class license will be able to operate:
 (a) CW only. (b) FM radiotelephony only. (c) AM radiotelephony only. (d) SSB radiotelephony only. (e) AM and CW.
15. Regarding the Rules and Regulations for the measurement of the frequencies of the emissions of amateur radio stations, select which of the following statements is correct:
 (a) The measurements must be taken on a periodic basis. (b) The measurements must be taken before each series of communications. (c) The measurements may be taken at any time. (d) The crystal-controlled frequency oscillator used in the transmitter may be used as a frequency standard. (e) The frequency can be checked by a separate crystal oscillator independent of the crystal oscillator used to control the frequency of the transmitter.
16. Radio waves travel through free space at the speed of:
 (a) 300,000,000 miles per second. (b) 30,000,000 miles per second. (c) 300,000,000 meters per second. (d) 30,000,000 meters per second. (e) 3,000,000 meters per second.
17. The relationship between the frequency and wavelength of a radio wave is:
 (a) The same. (b) Nonproportional. (c) Unrelated. (d) Directly proportional. (e) Inversely proportional.
18. Radio signals can travel over great distances because of the skip distances achieved by the reflection of the signals from the:
 (a) Ionized electrons. (b) Ground surface. (c) Skip angle. (d) Ionosphere. (e) Stratosphere.

19. During daylight hours the frequency band available to the Novice that is most likely to result in long distance communication is the:
(a) 80-m band. (b) 40-m band. (c) 20-m band. (d) 15-m band. (e) 10-m band.
20. During nighttime hours the frequency band available to the Novice that is most likely to result in long distance communication is the:
(a) 80-m band. (b) 40-m band. (c) 20-m band. (d) 15-m band. (e) 10-m band.
21. When transmitted by telegraphy $\overline{\text{SK}}$ means:
(a) Calling any station. (b) From. (c) Invitation to transmit. (d) End of message. (e) End of transmission.
22. When using the RST reporting system a signal that is perfectly readable, is extremely strong, and is a pure dc note would have a rating of:
(a) RST 599. (b) RST 999. (c) RST 555. (d) RST 579. (e) RST 979.
23. The Q-signal QRM means:
(a) The name of my station is   . (b) I am being interfered with. (c) I am being troubled by static. (d) Send more slowly. (e) Stop sending.
24. Q-signals become questions when followed by:
(a) $\overline{\text{AR}}$. (b) $\overline{\text{SK}}$. (c) $\overline{\text{AS}}$. (d) $\overline{\text{IMI}}$. (e) $\overline{\text{BT}}$.
25. Select the factor listed below that is *not* to be considered when selecting a transmitting frequency near one end of the authorized frequency band:
(a) Oscillator accuracy. (b) The authorized frequency of the band edge. (c) The error of the frequency checking instrument used to calibrate the oscillator. (d) The type of frequency checking instrument being used. (e) The bandwidth of the type of signal transmitted.
26. The symbol A1 is used to designate:
(a) Unmodulated carrier. (b) Radiotelephone. (c) Frequency-modulated radiotelegraphy using on-off keying. (d) Amplitude-modulated radiotelegraphy using on-off keying. (e) Phase-modulated radiotelegraphy using on-off keying.
27. EMF, or voltage, can also be referred to as:
(a) A difference in potential. (b) A difference in power. (c) A difference in force. (d) A difference in pressure. (e) A difference in levels.
28. Select the correct statement of units of measurement:
(a) EMF = volts; current = watts; power = amperes.
(b) EMF = amperes; current = watts; power = volts.
(c) EMF = watts; current = volts; power = amperes.
(d) EMF = watts; current = amperes; power = volts.
(e) EMF = volts; current = amperes; power = watts.
29. To convert alternating current to direct current we use a(n):
(a) Alternator. (b) Rectifier. (c) Triode. (d) Switch. (e) Amplifier.
30. A kilohertz equals:
(a) 10 Hz. (b) 100 Hz. (c) 1000 Hz. (d) 100,000 Hz. (e) 1,000,000 Hz.
31. The value of 8000 Hz is the fourth harmonic of the fundamental frequency of:
(a) 4000 Hz. (b) 400 Hz. (c) 200 Mz. (d) 2000 Hz. (e) 800 Hz.
32. The unit value for inductance is the:
(a) Oersted. (b) Ohm. (c) Henry. (d) Watt. (e) Joule.
33. The value of one-millionth of a farad is called a:
(a) Micromicrofarad. (b) Microfarad. (c) Farad. (d) Picofarad. (e) Millifarad.

34. A resistor whose value is 50 Ω has a voltage across it of 200 V; the current flowing through the resistor is:
(a) 10 A. (b) 1 A. (c) 0.25 A. (d) 40 A. (e) 4 A.
35. When the value of a resistive load equals the value of the internal resistance of a battery we have:
(a) No power delivered from the battery. (b) Maximum power delivered from the battery. (c) Partial power delivered from the battery. (d) Half power delivered from the battery. (e) A mismatch between battery and load.
36. The symbol F3 is used to designate:
(a) An unmodulated carrier. (b) Radiotelephony. (c) Amplitude-modulated radiotelegraphy using on-off keying. (d) Frequency-modulated radiotelephony. (e) Frequency-modulated radiotelephony using on-off keying.
37. A material in which the atomic electrons are tightly bound is a(n):
(a) Insulator. (b) Conductor. (c) Semiconductor. (d) Impurity. (e) Solid-state material.
38. Listed in the correct order are the names of the following symbols:
(a) Resistor, capacitor, inductor.
(b) Capacitor, inductor, resistor.
(c) Resistor, inductor, capacitor.
(d) Inductor, capacitor, resistor.
(e) Inductor, resistor, capacitor.
39. The following symbol is that of a:
(a) Semiconductor diode.
(b) Vacuum tube triode.
(c) Vacuum tube diode.
(d) NPN transistor.
(e) PNP transistor.

40. A half-wave antenna is one:
(a) Whose height above ground is a half-wavelength. (b) Whose length is one-half of a full wavelength. (c) Whose length is one-half of a half-wavelength. (d) Where the transmission line is connected at a half-wave in length. (e) Where the connecting transmission line is a half-wavelength long.
41. A half-wavelength antenna with the transmission line connected to the center of the antenna is a:
(a) Dipole antenna. (b) Bipole antenna. (c) Polar antenna. (d) Split antenna. (e) Quarter-wave antenna.
42. A half-wave antenna 125 feet long would be for use in the:
(a) 10-m band. (b) 15-m band. (c) 20-m band. (d) 40-m band. (e) 80-m band.
43. A transmission line is used to:
(a) Radiate radio waves. (b) Radiate radio frequency energy to the antenna. (c) Carry radio frequency energy to the antenna. (d) Match impedances between the antenna and transmitter. (e) Replace the antenna.
44. A transmission line consisting of a center conductor, a circular insulator, and a woven wire braid as an outer conductor is called a:
(a) Parallel-conductor cable. (b) Coaxial cable. (c) Twin-lead cable. (d) Circumferential cable. (e) Plastic cable.

45. The type of transmission line described in the question above has a typical impedance of:
(a) Between 100 and 700 Ω. (b) Between 400 and 600 Ω. (c) Approximately 300 Ω. (d) Either 50 or 72 Ω. (e) Either 72 or 150 Ω.
46. A multiband antenna is one that:
(a) Is multiple wavelengths long. (b) Is used on multiple amateur bands. (c) Multiplies the antenna frequency. (d) Is a multiple of a half-wavelength. (e) Multiplies the antenna length.
47. Select the one *incorrect* shock hazard precaution listed below:
(a) All metal chassis should be connected directly to ground. (b) A 3-wire grounded cable and connector should be used. (c) An ungrounded ac outlet should be used. (d) Power supplies should use bleeder resistors across filter capacitors. (e) External feed lines, and the antenna, should not be placed near power lines.
48. Operating simplex in the FM radiotelephony bands means you are:
(a) Operating directly with another station on one frequency. (b) Operating through a repeater on two frequencies. (c) Transmitting directly on one frequency and receiving directly on another. (d) Transmitting and receiving on one frequency through a repeater. (e) Transmitting and receiving on two bands through a repeater.
49. Multiplying the plate voltage by the plate current of the tube (or tubes) used in the final amplifier stage feeding the antenna will determine the:
(a) Plate power. (b) Radiated power. (c) Antenna power. (d) Input power. (e) Output power.
50. A special coupling circuit used between the output of the transmitter and the transmission line to have the output impedance of the transmitter match impedance of the transmission line is commonly referred to as a(n):
(a) Transcoupler. (b) Transmatch. (c) Antennamatch. (d) Impedamatch. (e) Impedacoupler.

Answers are on page 142.

FCC-Type Examination for a General Class License

1. The definition of portable operation is:
(a) Operation from a specific location shown on the station license. (b) Operation from an automobile that has stopped at a specific location. (c) Operation from a specific location other than that shown on the station license. (d) Removing and using the station equipment in an automobile. (e) Operation from an automobile that is in motion.
2. Third-party traffic from amateur operators in a foreign country is defined as:
(a) A radio communication to a third amateur operator other than the two amateur operators handling the message. (b) A radio communication that is to be delivered to three parties. (c) A radio communication that has been relayed

by three amateur operators. (d) A radio communication between amateur operators that is for anyone other than the amateur operators. (e) A radio communication between amateur operators that is restricted to authorized amateur operators.

3. Select the country below that does *not* have a third-party agreement with the United States:
(a) Argentina. (b) Colombia. (c) Ecuador. (d) Great Britain. (e) Israel.
4. A third party may participate in amateur radio communications provided that:
(a) The station has a valid amateur license. (b) A licensed radio amateur is acting as the control operator. (c) A licensed radio amateur is present during transmission. (d) The radio station is not a Novice licensed station. (e) Only phone communication is used.
5. Select the one method listed below by which an amateur radio station may *not* utilize one-way transmission:
(a) Third-party traffic. (b) Emergency communications. (c) Amateur radio information bulletins. (d) Net communications. (e) Code practice transmissions.
6. Select the one type of transmission listed below which is prohibited:
(a) Third-party messages. (b) Noncommercial messages for which no money has ben paid. (c) Television. (d) FM. (e) Music.
7. Select the one log entry below that is to be used on other than a daily basis:
(a) Date. (b) Transmitter input power. (c) Call sign of the station called. (d) Frequency used. (e) Type of emission used.
8. The log of a station must be preserved for a period of:
(a) No less than 1 year following the last date of entry. (b) No more than 1 year following the last date of entry. (c) No less than 2 years following the last date of entry. (d) No more than 2 years following the last date of entry. (e) For as long as the station is in operation.
9. Station operator privileges are determined by:
(a) The type of station license. (b) The portion of the amateur band selected for operation. (c) The type of control operator's license. (d) The type of equipment available. (e) The type of control point available.
10. Upon conviction, the maximum penalty for each day in which a violation of the FCC rules occurs can be for no more than:
(a) $50. (b) $100. (c) $500. (d) $1000. (e) $5000.
11. Notice of operation away from the authorized location of an amateur station is required when the period exceeds:
(a) 5 days. (b) 10 days. (c) 15 days. (d) 21 days. (e) 1 month.
12. Propagation characteristic of the VHF band is:
(a) Skip distances that vary with the frequency selected. (b) Skip distances that vary with the time of day selected. (c) Skip distances that vary with both the frequency and the time of day selected. (d) Line-of-sight distance. (e) Over-the-horizon distances.
13. Select the one answer that is *not* considered a good operating procedure:
(a) Use no more power than that required. (b) Use proper calling and answering procedures. (c) If available use a directive antenna. (d) When testing use a dummy antenna. (e) Use shielded coaxial cable for the transmission line.
14. A frequency-modulated radiotelephone signal is emission type:
(a) F1. (b) F3. (c) A1. (d) A3. (e) P5.

15. The desired range of audio frequencies that will convey best voice intelligibility is:
(a) 200–3000 Hz. (b) 20–3000 Hz. (c) 300–2000 Hz. (d) 500–2000 Hz. (e) 1000–3000 Hz.
16. When the amplitude of the speech-modulating signal directly varies the amplitude of the carrier signal, the type of emission is known as:
(a) PM. (b) FM. (c) TTY. (d) FSK. (e) AM.
17. Narrowband F3 emission permits the use of frequencies on either side of the carrier (sidebands) no greater than:
(a) 1 kHz. (b) 2 kHz. (c) 3 kHz. (d) 5 kHz. (e) 10 kHz.
18. Peak-envelope power (PEP) in an emission from an rf amplifier is:
(a) When the peak voltage does not exceed 0.707 of the peak current. (b) The value of 0.707 of the peak value of voltage and current. (c) When the peak power does not exceed 0.707 of the maximum power. (d) The value of 0.707 of the peak voltage. (e) The value of 0.707 of the peak current.
19. PEP input to the final amplifier stage supplying power to the antenna is determined by:
(a) Measurement with an rf voltmeter. (b) Measurement with an rf wattmeter. (c) Multiplying the metered value of plate voltage by the highest reading metered value of plate current while a modulating signal is being emitted. (d) Multiplying the metered value of plate voltage by the metered average value of plate current while a modulating signal is being emitted. (e) Multiplying the metered value of plate voltage by the metered value of plate current while an unmodulated signal is being emitted.
20. The PEP-to-average power ratio in a SSB transmitter is determined by:
(a) The voice-modulating signal waveform. (b) The voice-modulating signal amplitude. (c) The modulator power rating. (d) The adjustment of the balanced modulator. (e) The power output rating of the transmitter.
21. Typical PEP-to-average power ratios are:
(a) 10:1 and 20:1. (b) 20:1 and 30:1. (c) 1:1 and 2:1. (d) 2:1 and 3:1. (e) 3:1 and 4:1.
22. The term "SD ratio" in a SSB transmitter means:
(a) Signal-to-decibel ratio. (b) Signal-to-distortion ratio. (c) Sensitivity-to-distortion ratio. (d) Sensitivity-to-deviation ratio. (e) Signal-to-Deviation ratio.
23. RFI stands for:
(a) Radio free of interference. (b) Radio frequency image. (c) Radio frequency interference. (d) Repeater frequency image. (e) Repeater frequency interference.
24. The total resistance of the following circuit is:
(a) 230 kΩ.
(b) 460 kΩ.
(c) 180 kΩ.
(d) 280 kΩ.
(e) 115 kΩ.

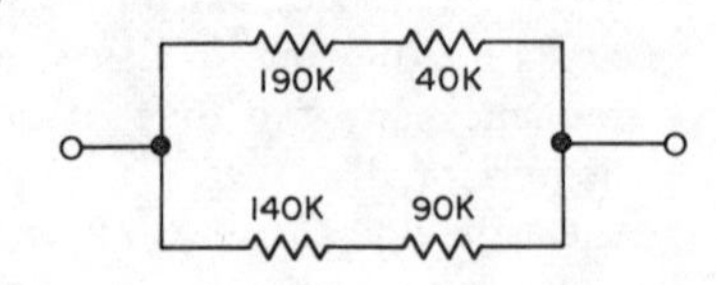

25. Voltage across series-connected capacitors:
(a) Divides evenly across all capacitors. (b) Divides across each capacitor as determined by the total value of capacity. (c) Divides across each capacitor as

determined by the value of the capacitor. (d) Divides across each capacitor as determined by the value of the voltage applied. (e) Divides across each capacitor as determined by the value of the capacitor's working voltage rating.

26. Inductive reactance is:
(a) The opposition of an inductor to alternating current. (b) The opposition of an inductor to direct current. (c) The value in ohms of the resistance of the inductor winding. (d) The reaction of the inductor winding to the expanding magnetic field. (e) The reaction of the inductor winding to the contracting magnetic field.
27. Capacitive reactance is:
(a) The opposition of a capacitor to alternating current. (b) The opposition of a capacitor to direct current. (c) The value in ohms of the resistance of the capacitor plates. (d) The value in ohms of the insulating material between the plates. (e) The reaction of the capacitor to the electric field between the plates.
28. Like reactances in parallel:
(a) Combine and are added together for the total value. (b) Oppose and the resulting value is the value of the lowest reactance. (c) Combine and the total value is less than that of the lowest value reactance. (d) Combine and the total value is that of the highest value reactance. (e) Combine and the total value is more than that of the highest value reactance.
29. The 470-Ω cathode resistor of an amplifier tube with a plate current of 30 mA has a voltage drop across it of:
(a) 1.41 V. (b) 14.1 V. (c) 141 V. (d) 15.6 V. (e) 5 V.
30. A decibel represents:
(a) One-tenth of a linear power ratio. (b) One-tenth of a unit of a power ratio. (c) A logarithmic value of one-tenth of a power ratio measurement. (d) A linear value of one-tenth of a power ratio measurement. (e) A linear value of a power ratio measurement.
31. A series-resonant circuit offers:
(a) Maximum current flow in the tank circuit. (b) Minimum current flow in the tank circuit. (c) Maximum current flow and maximum impedance. (d) Minimum current flow and maximum impedance. (e) Maximum current flow and minimum impedance.
32. A parallel-resonant circuit having a reactance of 480 Ω and a resistance of 16 Ω has a Q of:
(a) 7680. (b) 464. (c) 496. (d) 30. (e) 16.
33. A transformer having 200 turns in the primary and 600 turns in the secondary having 250 V applied to the primary will result in the secondary winding having a voltage value of:
(a) 200 V. (b) 60 V. (c) 250 V. (d) 83.3 V. (e) 750 V.
34. To have an amplifier oscillate, we must supply:
(a) In-phase feedback. (b) Out-of-phase feedback. (c) In-phase signals. (d) Out-of-phase signals. (e) Negative feedback.
35. Select the one statement below that is *not* a correct method of minimizing harmonic generation:
(a) Properly tuning the resonant circuits. (b) Using a minimum value of grid drive signal. (c) Using a maximum value of plate current. (d) Using the correct value of grid bias voltage. (e) Use linear amplifier circuits.

36. Power amplification of radio frequency signals is best done by using an amplifier that is:
(a) Class A. (b) Class AB. (c) Class AB^1. (d) Class B. (e) Class C.
37. To prevent undesired feedback in an amplifier circuit, we use:
(a) Neutralization. (b) Parasitic feedback. (c) Positive feedback. (d) Plate voltage feedback. (e) Grid voltage feedback.
38. Select the answer that lists in the correct order the three basic circuits shown below:

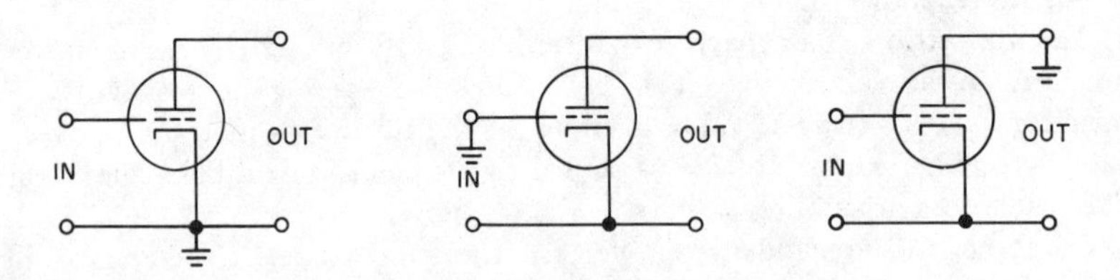

(a) Grounded cathode, grounded plate, grounded grid.
(b) Grounded cathode, grounded grid, grounded plate.
(c) Grounded grid, grounded cathode, grounded plate.
(d) Grounded grid, grounded plate, grounded cathode.
(e) Grounded plate, grounded grid, grounded cathode.
39. Semiconductor diodes are constructed of:
(a) P-type semiconductor material. (b) N-type semiconductor material. (c) A junction of N-type semiconductor materials. (d) A junction of P-type semiconductor materials. (e) A junction of N- and P-type semiconductor materials.
40. Semiconductor transistors are constructed of:
(a) P-type semiconductor material. (b) N-type semiconductor material. (c) A junction of P- and N-type semiconductor materials. (d) Two junctions of N- and P-type materials. (e) Two junctions of N- or P-type materials.
41. In the symbol of a transistor shown below the correct nomenclature for each element is:
(a) 1, Base; 2, emitter; 3, collector.
(b) 1, Base; 2, collector; 3, emitter.
(c) 1, Emitter; 2, collector; 3, base.
(d) 1, Emitter; 2, base; 3, collector.
(e) 1, Collector; 2, base; 3, emitter.

1 2 3

42. Due to the construction of the electrolytic capacitor, it can:
(a) Be used with any value of voltage. (b) Be used only with alternating current. (c) Be used only with direct current. (d) Be used only with very low values of voltage. (e) Be used with any value of current.
43. A toroidal inductor is:
(a) A very-high-impedance coil winding. (b) A very-low-impedance coil winding. (c) A pie-shaped coil winding. (d) A doughnut-shaped coil winding. (e) A round-shaped coil winding.
44. A half-wave antenna is one:
(a) Whose height above ground is one half-wavelength. (b) Whose length is one-half of a full wavelength. (c) Whose length is one-half of a half-wavelength.

(d) Where the transmission line is connected at a half-wave in length. (e) Where the connecting transmission line is one half-wavelength long.

45. Radiation from a quarter-wave ground plane antenna is:
(a) Directional. (b) Omnidirectional. (c) Bidirectional. (d) Vertical only. (e) Horizontal only.
46. For operation on several HF bands, a center-fed horizontal nonresonant antenna fed by a parallel-conductor transmission line should use between the output of the transmitter and the transmission line a(n):
(a) Transmatch. (b) Transcoupler. (c) Antennamatch. (d) Impedamatch. (e) Impedacoupler.
47. Select the correct formula for use in finding the standing wave ratio (SWR) using the incident or reflected voltage:
(a) $\text{SWR} = \dfrac{\text{max}}{\text{min}}$. (b) $\text{SWR} = \dfrac{E_{\text{max}}}{I_{\text{min}}}$. (c) $\text{SWR} = \dfrac{I_{\text{max}}}{E_{\text{max}}}$. (d) $\text{SWR} = \dfrac{I_{\text{min}}}{E_{\text{min}}}$.
(e) $\text{SWR} = \dfrac{E_{\text{max}}}{E_{\text{min}}}$.
48. The major characteristics that determine the impedance of a parallel-conductor transmission line are:
(a) The number of conductors and the distance between them. (b) The number of conductors and the diameter of the conductors. (c) The distance between the conductors and the diameter of the conductors. (d) The type of conductors and their diameter. (e) The type of conductors and the distance between them.
49. The one test instrument capable of giving more accurate information for the proper adjustment of a radiotelephone transmitter is the:
(a) Voltmeter. (b) Vacuum-tube voltmeter. (c) Multimeter. (d) Oscilloscope. (e) Signal generator.
50. Shown below is a block diagram of a filter system SSB transmitter. The unmarked block is the:

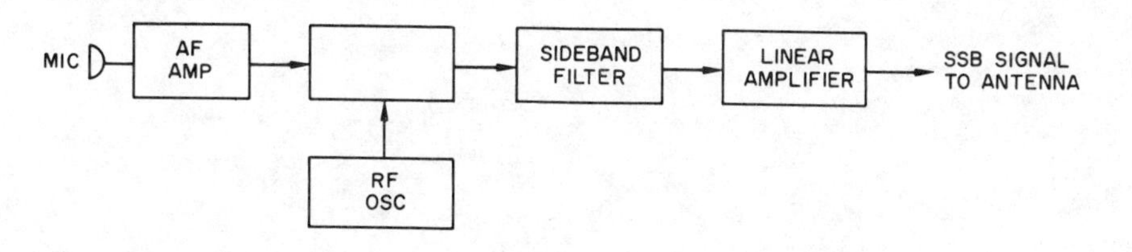

(a) Balanced detector. (b) Balanced modulator. (c) Sideband modulator. (d) Sideband detector. (e) Linear modulator.

Answers are on page 142.

Answers to FCC-Type Examination for a Novice Class License

1. (a)	11. (a)	21. (e)	31. (d)	41. (a)
2. (c)	12. (c)	22. (a)	32. (c)	42. (e)
3. (e)	13. (b)	23. (b)	33. (b)	43. (c)
4. (d)	14. (b)	24. (d)	34. (e)	44. (b)
5. (b)	15. (a)	25. (d)	35. (b)	45. (d)
6. (d)	16. (c)	26. (d)	36. (d)	46. (b)
7. (c)	17. (e)	27. (a)	37. (a)	47. (c)
8. (a)	18. (d)	28. (e)	38. (c)	48. (a)
9. (b)	19. (d)	29. (b)	39. (c)	49. (d)
10. (e)	20. (b)	30. (c)	40. (b)	50. (b)

Answers to FCC-Type Examination for a General Class License

1. (c)	11. (c)	21. (d)	31. (e)	41. (b)
2. (d)	12. (d)	22. (b)	32. (d)	42. (c)
3. (d)	13. (e)	23. (c)	33. (e)	43. (d)
4. (b)	14. (b)	24. (e)	34. (a)	44. (b)
5. (a)	15. (a)	25. (c)	35. (c)	45. (b)
6. (e)	16. (e)	26. (a)	36. (e)	46. (a)
7. (b)	17. (c)	27. (a)	37. (a)	47. (e)
8. (a)	18. (b)	28. (c)	38. (b)	48. (c)
9. (c)	19. (c)	29. (b)	39. (e)	49. (d)
10. (c)	20. (a)	30. (c)	40. (d)	50. (b)

Appendix II

FEDERAL REGULATIONS

Excerpts from the Communications Act of 1934, as Amended

BEING AN ACT TO PROVIDE FOR THE REGULATION OF INTERSTATE AND FOREIGN COMMUNICATION BY WIRE OR RADIO, AND FOR OTHER PURPOSES

Be it enacted by the Senate and House of Representatives of the United States of America in Congress assembled,

SEC. 1. For the purpose of regulating interstate and foreign commerce in communication by wire and radio so as to make available, so far as possible, to all the people of the United States a rapid, efficient, Nation-wide, and world-wide wire and radio communication service with adequate facilities at reasonable charges, for the purpose of the national defense, for the purpose of promoting safety of life and property through the use of wire and radio communication, and for the purpose of securing a more effective execution of this policy by centralizing authority heretofore granted by law to several agencies and by granting additional authority with respect to interstate and foreign commerce in wire and radio communication, there is hereby created a commission to be known as the "Federal Communications Commission," which shall be constituted as hereinafter provided, and which shall execute and enforce the provisions of this Act.

SEC. 2. (a) The provisions of this Act shall apply to all interstate and foreign communication by wire or radio and all interstate and foreign transmission of energy by radio, which originates and/or is received within the United States, and to all persons engaged within the United States in such communication or such transmission of energy by radio, and to the licensing and regulating of all radio stations as hereinafter provided; but it shall not apply to persons engaged in wire or radio communication or transmission in the Canal Zone, or to wire or radio communication or transmission wholly within the Canal Zone.

SEC. 4. (a) The Federal Communications Commission (in this Act referred to as the "Commission") shall be composed of seven commissioners appointed by the President, by and with the advice and consent of the Senate, one of whom the President shall designate as chairman.

(i) The Commission may perform any and all acts, make such rules and regulations, and issue such orders, not inconsistent with this Act, as may be necessary in the execution of its functions.

SEC. 301. It is the purpose of this Act, among other things, to maintain the control of the United States over all the channels of interstate and foreign radio transmission; and to provide for the use of such channels, but not the ownership thereof, by persons for limited periods of time, under licenses granted by Federal authority, and no such license shall be construed to create any right, beyond the terms, conditions, and periods of the license. No person shall use or operate any apparatus for the transmission of energy or communications or signals by radio . . . except under and in accordance with this Act and with a license in that behalf granted under the provisions of this Act.

SEC. 303. Except as otherwise provided in this Act, the Commission from time to time, as public convenience, interest, or necessity requires shall—

(a) Classify radio stations;

(b) Prescribe the nature of the service to be rendered by each class of licensed stations and each station within any class;

(c) Assign bands of frequencies to the various classes of stations, and assign frequencies for each individual station and determine the power which each station shall use and the time during which it may operate;

(d) Determine the location of classes of stations or individual stations;

(e) Regulate the kind of apparatus to be used with respect to its external effects and the purity and sharpness of the emissions from each station and from the apparatus therein;

(f) Make such regulations not inconsistent with law as it may deem necessary to prevent interference between stations and to carry out the provisions of this Act: *Provided, however,* that changes in the frequencies, authorized power, or in the times of operation of any station, shall not be made without the consent of the station licensee unless, after a public hearing, the Commission shall determine that such changes will promote public convenience or interest or will serve public necessity, or the provisions of this Act will be more fully complied with;

(g) Study new uses for radio, provide for experimental uses of frequencies, and generally encourage the larger and more effective use of radio in the public interest;

(h) Have authority to establish areas or zones to be served by any station;

(i) Have authority to make special regulations applicable to radio stations engaged in chain broadcasting;

(j) Have authority to make general rules and regulations requiring stations to keep such records of programs, transmissions of energy, communications, or signals as it may deem desirable;

(k) Have authority to exclude from the requirements of any regulations in whole or in part any radio station upon railroad rolling stock, or to modify such regulations in its discretion;

(l) Have authority to prescribe the qualifications of station operators, to classify them according to the duties to be performed, to fix the forms of such licensees, and to issue them to such citizens or nationals of the United States as the Commission finds qualified, except that in issuing licenses for the operation of radio stations on aircraft the Commission may, if it finds that the public interest will be served thereby, waive the requirement of citizenship in the case of persons holding United States pilot certificates or in the case of persons holding foreign aircraft pilot certificates which are valid in the United States on the basis of reciprocal agreements entered into with foreign governments;

(m) (1) Have authority to suspend the license of any operator upon proof sufficient to satisfy the Commission that the licensee—

(A) Has violated any provision of any Act, treaty, or convention binding on the United States, which the Commission is authorized to administer, or any regulation made by the Commission under any such Act, treaty, or convention; or

(B) Has failed to carry out a lawful order of the master or person lawfully in charge of the ship or aircraft on which he is employed; or

(C) Has willfully damaged or permitted radio apparatus or installations to be damaged; or

(D) Has transmitted superfluous radio communications or signals or communications containing profane or obscene words, language, or meaning, or has knowingly transmitted—

(1) False or deceptive signals or communications, or

(2) A call signal or letter which has not been assigned by proper authority to the station he is operating; or

(E) Has willfully or maliciously interfered with any other radio communications or signals; or

(F) Has obtained or attempted to obtain, or has assisted another to obtain or attempted to obtain, an operator's license by fraudulent means.

(n) Have authority to inspect all radio installations associated with stations required to be licensed by any Act or which are subject to the provisions of any Act, treaty, or convention binding on the United States, to ascertain whether in construction, installation, and operation they conform to the requirements of the rules and regulations of the Commission, the provisions of any Act, the terms of any treaty or convention binding on the United States, and the conditions of the license or other instrument of authorization under which they are constructed, installed, or operated.

(o) Have authority to designate call letters of all stations;

(p) Have authority to cause to be published such call letters and such other announcements and data as in the judgment of the Commission may be required for the efficient operation of radio stations subject to the jurisdiction of the United States and for the proper enforcement of this Act:

(q) Have authority to require the painting and/or illumination of radio towers if and when in its judgment such towers constitute, or there is a reasonable possibility that they may constitute, a menace to air navigation.

(r) Make such rules and regulations and prescribe such restrictions and conditions, not inconsistent with law, as may be necessary to carry out the provisions of this Act, or any international radio or wire communications treaty or convention, or regulations annexed thereto, including any treaty or convention insofar as it relates to the use of radio, to which the United States is or may hereafter become a party.

SEC. 318. The actual operation of all transmitting apparatus in any radio station for which a station license is required by this Act shall be carried on only by a person holding an operator's license issued hereunder, and no person shall operate any such apparatus in such station except under and in accordance with an operator's license issued to him by the Commission: *Provided, however,* That the Commission if it shall find that the public interest, convenience, or necessity will be served thereby may waive or modify the foregoing provisions of this section for the operation of any station except (1) stations for which licensed operators are required by international agreement, (2) stations for which licensed operators are required for safety purposes, (3) stations engaged in broadcasting (other than those engaged solely in the function of rebroadcasting the signals of television broadcast stations), and (4) stations operated as common carriers on frequencies below thirty thousand kilocycles: *Provided further,* That the Commission shall have power to make special regulations governing the granting of licenses for the use of automatic radio devices and for the operation of such devices.

SEC. 321. (a) The transmitting set in a radio station on ship-board may be adjusted in such a manner as to produce a maximum of radiation, irrespective of the amount of interference which may thus be caused, when such station is sending radio communications or signals of distress and radio communications relating thereto.

(b) All radio stations, including Government stations and stations on board foreign vessels when within the territorial waters of the United States, shall give absolute priority to radio communications or signals relating to ships in distress; shall cease all sending on frequencies which will interfere with hearing a radio communication or signal of distress, and, except when engaged in answering or aiding the ship in distress, shall refrain from sending any radio communications or signals until there is assurance that no interference will be caused with the radio communications or signals relating thereto, and shall assist the vessel in distress, so far as possible, by complying with its instructions.

SEC. 324. In all circumstances, except in case of radio communications or signals relating to vessels in distress, all radio stations, including those owned and operated by the United States, shall use the minimum amount of power necessary to carry out the communication desired.

SEC. 325. (a) No person within the jurisdiction of the United States shall knowingly utter or transmit, or cause to be uttered or transmitted, any false or fraudulent signal of distress, or communication relating thereto, nor shall any broadcasting station rebroadcast the program or any part thereof of another broadcasting station without the express authority of the originating station.

SEC. 326. Nothing in this Act shall be understood or construed to give the Commission the power of censorship over the radio communications or signals transmitted by any radio station, and no regulation or condition shall be promulgated or fixed by the Commission which shall interfere with the right of free speech by means of radio communication.

SEC. 501. Any person who willfully and knowingly does or causes or suffers to be done any act, matter, or thing, in this Act prohibited or declared to be unlawful, or who willfully or knowingly omits or fails to do any act, matter, or thing in this Act required to be done, or willfully and knowingly causes or suffers such omission or failure, shall, upon conviction thereof, be punished for such offense, for which no penalty (other than a forfeiture) is provided in this Act, by a fine of not more than $10,000 or by imprisonment for a term not exceeding one year, or both; except that any person, having been once convicted of an offense punishable under this section, who is subsequently convicted of violating any provision of this Act punishable under this section, shall be punished by a fine of not more than $10,000 or by imprisonment for a term not exceeding two years, or both.

SEC. 502. Any person who willfully and knowingly violates any rule, regulation, restriction, or condition made or imposed by the Commission under authority of this Act, or any rule, regulation, restriction, or condition made or imposed by any international radio or wire communications treaty or convention, or regulations annexed thereto, to which the United States is or may hereafter become a party, shall, in addition to any other penalties provided by law, be punished, upon conviction thereof, by a fine of not more than $500 for each and every day during which such offense occurs.

SEC. 605. No person receiving or assisting in receiving, or transmitting, or assisting in transmitting, any interstate or foreign communication by wire or radio shall divulge or publish the existence, contents, substance, purport, effect, or meaning thereof, except through authorized channels of transmission or reception, to any person other than the addressee, his agent, or attorney, or to a person employed or authorized to forward such communication to its destination, or to proper accounting or distributing officers of the various communicating centers over which the communication may be passed, or to the master of a ship under whom he is serving, or in response to a subpena issued by a court of competent jurisdiction, or on demand of other lawful authority; and no person not being authorized by the sender shall intercept any communication and divulge or publish the existence, contents, substance, purport, effect, or meaning of such intercepted communication to any person; and no person not being entitled thereto shall receive or assist in receiving any interstate or foreign communication by wire or radio and use the same or any information therein contained for his own benefit or for the benefit of another not entitled thereto; and no person having received such intercepted communication or having become acquainted with the contents, substance, purport, effect, or meaning of the same or any part thereof, knowing that such information was so obtained, shall divulge or publish the existence, contents, substance, purport, effect, or meaning of the same or any part thereof, or use the same or any information therein contained for his own benefit or for the benefit of another not entitled thereto: *Provided,* That this section shall not apply to the receiving, divulging, publishing, or utilizing the contents of any radio communication broadcast, or transmitted by amateurs or others for the use of the general public, or relating to ships in distress.

SEC. 606. (a) During the continuance of a war in which the United States is engaged, the President is authorized, if he finds it necessary for the national defense and security, to direct that such communications as in his judgment may be essential to the national defense and security shall have preference or priority with any carrier subject to this Act. He may give these directions at and for such times as he may determine, and may modify, change, suspend, or annul them and for any such purpose he is hereby authorized to issue orders directly, or through such person or persons as he designates for the purpose, or through the Commission. Any carrier complying with any such order or direction for preference or priority herein authorized shall be exempt from any and all provisions in existing law imposing civil or criminal penalties, obligations, or liabilities upon carriers by reason of giving preference or priority in compliance with such order or direction.

(b) It shall be unlawful for any person during any war in which the United States is engaged to knowingly or willfully, by physical force or intimidation by threats of physical force, obstruct or retard or aid in obstructing or retarding interstate or foreign communication by radio or wire. The President is hereby authorized, whenever in his judgment the public interest requires, to employ the armed forces of the United States to prevent any such obstruction or retardation of communication: *Provided,* That nothing in this section shall be construed to repeal, modify, or affect either section 6 or section 20 of the Act entitled "An Act to supplement existing laws against unlawful restraints and monopolies, and for other purposes", approved October 15, 1914.

(c) Upon proclamation by the President that there exists war or a threat of war, or a state of public peril or disaster or other national emergency, or in order to preserve the neutrality of the United States, the President, if he deems it necessary in the interest of national security, or defense, may suspend or amend, for such time as he may see fit, the rules and regulations applicable to any or all stations or devices capable of emitting electromagnetic radiations within the jurisdiction of the United States as prescribed by the Commission, and may cause the closing of any station for radio communication, or any device capable of emitting electromagnetic radiations between 10 kilocycles and 100,000 megacycles, which is suitable for use as a navigational aid beyond five miles, and the removal therefrom of its apparatus and equipment, or he may authorize the use or control of any such station or device and/or its apparatus and equipment, by any department of the Government under such regulations as he may prescribe upon just compensation to the owners. The authority granted to the President, under this subsection, to cause the closing of any station or device, and the removal therefrom of its apparatus and equipment, or to authorize the use or control of any station or device and/or its apparatus and equipment, may be exercised in the Canal Zone.

Excerpts from the FCC Rules and Regulations: Part 97, *Amateur Radio Service*

SUBPART A—GENERAL

§ 97.1 Basis and purpose.

The rules and regulations in this part are designed to provide an amateur radio service having a fundamental purpose as expressed in the following principles:

(a) Recognition and enhancement of the value of the amateur service to the public as a voluntary noncommercial communication service, particularly with respect to providing emergency communications.

(b) Continuation and extension of the amateur's proven ability to contribute to the advancement of the radio art.

(c) Encouragement and improvement of the amateur radio service through rules which provide for advancing skills in both the communication and technical phases of the art.

(d) Expansion of the existing reservoir within the amateur radio service of trained operators, technicians, and electronics experts.

(e) Continuation and extension of the amateur's unique ability to enhance international good will.

§ 97.3 Definitions.

(a) *Amateur radio service.* A radio communication service of self-training, intercommunication, and technical investigation carried on by amateur radio operators.

(b) *Amateur radio communication.* Noncommercial radio communication by or among amateur radio stations solely with a personal aim and without pecuniary or business interest.

(c) *Amateur radio operator.* A person interested in radio technique solely with a personal aim and without pecuniary interest, holding a valid Federal Communications Commission license to operate amateur radio stations.

(d) *Amateur radio license.* The instrument of authorization issued by the Federal Communications Commission comprised of a station license, and in the case of the primary station, also incorporating an operator license.

Operator license. The instrument of operator authorization including the class of operator privileges.

Station license. The instrument of authorization for a radio station in the amateur radio service.

(e) *Amateur radio station.* A station licensed in the amateur radio service embracing necessary apparatus at a particular location used for amateur radio communication.

(f) *Primary station.* The principal amateur radio station at a specific land location shown on the station license.

(g) *Military recreation station.* An amateur radio station licensed to the person in charge of a station at a land location provided for the recreational use of amateur radio operators, under military auspices of the Armed Forces of the United States.

(h) *Club station.* A separate amateur radio station for use by the members of a bona fide amateur radio society and licensed to an amateur radio operator acting as the station trustee for the society.

(i) *Additional station.* Any amateur radio station licensed to an amateur radio operator normally for a specific land location other than the primary station, may be one or more of the following:

Secondary station. Station licensed for a land location other than the primary station location, i.e., for use at a subordinate location such as an office, vacation home, etc.

Control station. Station licensed to conduct remote control of another amateur radio station.

Auxiliary link station. Station, other than a repeater station, at a specific land location licensed only for the purpose of automatically relaying radio signals from that location to another specific land location.

Repeater station. Station licensed to automatically retransmit the radio signals of other amateur radio stations for the purpose of extending their intracommunity radio communication range.

(j) *Space radio station.* An amateur radio station located on an object which is beyond, is intended to go beyond, or has been beyond the major portion of the earth's atmosphere. (Regulations governing this type of station have not yet been adopted and all applications will be considered on an individual basis.)

(k) *Terrestrial location.* Any point within the major portion of the earth's atmosphere, including aeronautical, land, and maritime locations.

(l) *Space location.* [Reserved]

(m) *Amateur radio operation.* Amateur radio communication conducted by an amateur radio operator from an amateur radio station. May include one or more of the following:

Fixed operation. Radio communication conducted from the specific geographical land location shown on the station license.

Portable operation. Radio communication conducted from a specific geographical location other than that shown on the station license.

Mobile operation. Radio communication conducted while in motion or during halts at unspecified locations.

(n) *Remote control.* Control of transmitting apparatus of an amateur radio station from a position other than one at which the transmitter is located and immediately accessible, except that direct mechanical control, or direct electrical control by wired connections, of an amateur radio transmitter from a point located on board any aircraft, vessel, vehicle, or on the same premises on which the transmitter is located, shall not be considered remote control within the meaning of this definition.

(o) *Control link.* Apparatus for effecting remote control between a control point and a remotely controlled station.

(p) *Control operator.* An amateur radio operator designated by the licensee of an amateur radio station to also be responsible for the emissions from that station.

(q) *Control point.* The operating position of an amateur radio station where the control operator function is performed.

(r) *Antenna structures.* Antenna structures include the radiating system, its supporting structures, and any appurtenances mounted thereon.

(s) *Antenna height above average terrain.* The height of the center of radiation of an antenna above an averaged value of the elevation above sea level for the surrounding terrain.

(t) *Transmitter.* Apparatus for converting electrical energy received from a source into radio-frequency electromagnetic energy capable of being radiated.

(u) *Effective radiated power.* The product of the radio-frequency power, expressed in watts, delivered to an antenna, and the relative gain of the antenna over that of a half-wave dipole antenna.

(v) *System network diagram.* A diagram showing each station and its relationship to the other stations in a network of stations, and to the control point(s).

(w) *Third-party traffic.* Amateur radio communication by or under the supervision of the control operator at an amateur radio station to another amateur radio station on behalf of anyone other than the control operator.

(x) *Emergency communication.* Any amateur radio communication directly relating to the immediate safety of life of individuals or the immediate protection of property.

SUBPART B—AMATEUR OPERATOR AND STATION LICENSES

Operator Licenses

§ 97.5 Classes of operator licenses.

Amateur extra class.
Advanced class (previously class A).
General class (previously class B).
Conditional class (previously class C).
Technician class.
Novice class.

§ 97.7 Privileges of operator licenses.

(a) *Amateur Extra Class and Advanced Class.* All authorized amateur privileges including exclusive frequency operating authority in accordance with the following table:

Frequencies	*Class of license authorized*
3500–3525 kHz	Amateur Extra Only.
3775–3800 kHz	
7000–7025 kHz	
14,000–14,025 kHz	
21,000–21,025 kHz	
21,250–21,270 kHz	
3800–3890 kHz	Amateur Extra and Advanced.
7150–7225 kHz	
14,200–14,275 kHz	
21,270–21,350 kHz	
50–50.1 MHz	

(b) *General Class and Conditional Class.* All authorized amateur privileges except those exclusive frequency operating privileges which are reserved to the Advanced Class and/or the Amateur Extra Class.

(c) *Technician class.* All authorized amateur privileges on the frequencies 50.1–54.0 MHz and 145–148 MHz and in the amateur frequency bands above 220 MHz.

(d) *Novice class.* Those amateur privileges designated and limited as follows:

(1) The power input to the transmitter final amplifying stage supplying radio frequency energy to the antenna shall not exceed 75 watts, exclusive of power for heating the cathode of a vacuum tube(s).

(2) Radio telegraphy is authorized in the frequency bands 3700–3750 kHz, 7100–7150 kHz (7050–7075 kHz when the terrestrial location of the station is not within Region 2), 21,100–21,200 kHz, and 28,100–28,200 kHz, using only Type A–1 emission.

§ 97.9 Eligibility for new operator license.

Persons are eligible to apply for the various classes of amateur operator licenses as follows:

(a) *Amateur extra class.* Any citizen or national of the United States who either (1) any time prior to receipt of his application by the Commission has held for at least 1 year an amateur operator license of other than the novice or technician class, issued by any agency of the U.S. Government, or submits proof that he held for a period of 1 year an amateur operator license at least equivalent to a general class license issued by a foreign government, or (2) submits evidence of having held a valid amateur radio station or operator license issued by any agency of the U.S. Government during or prior to April 1917.

(b) *Advanced Class.* Any citizen or national of the United States.

(c) *General class.* Any citizen or national of the United States.

(d) *Conditional class.* Any citizen or national of the United States:

(1) Whose actual residence and amateur station location are more than 175 miles airline distance from the nearest location at which examinations are held at intervals of not more than 6 months for General Class amateur operator licenses.

(2) Who is shown by physician's certificate to be unable to appear for examination because of protracted disability.

(3) Who is shown by certificate of the commanding officer to be in the armed forces of the United States at any Army, Navy, Air Force, or Coast Guard station and, for that reason, to be unable to appear for examination at the time and place designated by the Commission.

(4) Who furnishes sufficient evidence, at the time of filing, of temporary residence for a continuous period of at least 12 months outside the continental limits of the United States, its territories or possessions, irrespective of other provisions of this paragraph.

(e) *Technician class.* Any citizen or national of the United States.

(f) *Novice Class.* Any citizen or national of the United States, except a person who holds, or who has held within the 12-month period prior to the date of receipt of his application, a Commission-issued amateur radio license. The Novice Class license may not be concurrently held with any other class of amateur radio license.

§ 97.11 Application for operator license.

(a) An application (FCC Form 610) for a new operator license, including an application for change in operating privileges, which will require an examination supervised by Commission personnel at a regular Commission examining office shall be submitted to such office in advance of or at the time of the examination, except that, whenever an examination is to be taken at a designated examination point away from a Commission office, the application, together with the necessary filing fee should be submitted in advance of the examination date to the office which has jurisdiction over the examination point involved.

(b) An application (FCC Form 610) for a new operator license, including an application for change in operating privileges, which requests an examination supervised by a volunteer examiner under the provisions of § 97.29(b), shall be submitted to the Commission's office at Gettysburg, Pennsylvania, 17325. The application shall be accompanied by any necessary filing fee and by a request for the written examination material (see § 97.29(b)).

(c) An application (FCC Form 610) for renewal and/or modification of license when no change in operating privileges is involved shall be submitted, together with any necessary filing fee, to the Commission's office at Gettysburg, Pennsylvania, 17325.

§ 97.13 Renewal or modification of operator license.

(a) An amateur operator license, except the Novice Class, may be renewed upon proper application in which it is stated that the applicant has lawfully accumulated, at an amateur station licensed by the Commission, a minimum total of either 2 hours operating time during the last 3 months or 5 hours operating time during the last 12 months of the license term. Such operating time, for the purpose of renewal, shall be counted as the total of all that time between the entries in the station log showing the beginning and end of transmissions as required in § 97.103(a), both during single transmissions and during a sequence of transmissions. The application shall, in addition to the foregoing, include a statement that the applicant can send by hand key, i.e., straight key or any other type of hand operated key such as a semi-automatic or electronic key, and receive by ear, in plain language, messages in the International Morse Code at a speed of not less than that which is required in qualifying for an original license of the class being renewed.

NOTE: Until further order of the Commission, the showing that the applicant actually operated an amateur radio station or stations for the periods of time specified in § 97.13 will not be required in cases where it is shown that the applicant was unable to conduct such operation because he was on active duty overseas in the armed forces of the United States or was duly enrolled as an employee of an agency of the Federal Government and in the course of such employment was on duty in a foreign country continuously during the last year of the license term: *Provided,* That any such employee of the Federal Government shall submit with his application for renewal of license a statement signed by his agency head, or the chief of the Bureau or Division in which he is employed attesting to such employment.

(b) The Novice Class license will not be renewed.

(c) The applicant shall qualify for a new license by examination if the requirements of this section are not fulfilled.

(d) Application for renewed and/or modification of an amateur operator license shall be submitted on FCC Form 610 and shall be accompanied by the applicant's license. Application for renewal of unexpired licenses must be made during the license term and should be filed within 90 days but not later than 30 days prior to the end of the license term. In any case in which the licensee has, in accordance with the provisions of this chapter, made timely and sufficient application for renewal of an unexpired license, no license with reference to any activity of a continuing nature shall expire until such application shall have been finally determined.

(e) If a license is allowed to expire, application for renewal may be made during a period of grace of one year after the expiration date. During this one year period of grace, an expired license is not valid. A license renewed during the grace period will be dated currently and will not be backdated to the date of its expiration. Application for renewal shall be submitted on FCC Form 610 and shall be accompanied by the applicant's expired license.

(f) When the name of a licensee is changed or when the mailing address is changed a formal application for modification of license is not required. However, the licensee shall notify the Commission promptly of these changes. The notice, which may be in letter form, shall contain the name and address of the licensee as they appear in the Commission's records, the new name and/or address, as the case may be, the radio station call sign and class of operator license. The notice shall be sent to Federal Communications Commission, Gettysburg, Pa., 17325, and a copy shall be kept by the licensee until a new license is issued.

OPERATOR LICENSE EXAMINATIONS

§ 97.19 When examination is required.

Examination is required for the issuance of a new amateur operator license, and for a change in class of operating privileges. Credit may be given, however, for certain elements of examination as provided in § 97.25.

§ 97.21 Examination elements.

Examinations for amateur operator privileges will comprise one or more of the following examination elements:

(a) Element 1(A): Beginner's code test at five (5) words per minute;

(b) Element 1(B): General code test at thirteen (13) words per minute;

(c) Element 1(C): Expert's code test at twenty (20) words per minute;

(d) Element 2: Basic law comprising rules and regulations essential to beginners' operation, including sufficient elementary radio theory for the understanding of those rules;

(e) Element 3: General amateur practice and regulations involving radio operation and apparatus and provisions of treaties, statutes, and rules affecting amateur stations and operators;

(f) Element 4(A): Intermediate amateur practice involving intermediate level radio theory and operation as applicable to modern amateur techniques, including, but not limited to, radiotelephony and radiotelegraphy;

(g) Element 4(B): Advanced amateur practice involving advanced radio theory and operation as applicable to modern amateur techniques, including, but not limited to, radiotelephony, radiotelegraphy, and transmissions of energy for measurements and observations applied to propagation, for the radio control of remote objects and for similar experimental purposes.

§ 97.23 Examination requirements.

Applicants for original licenses will be required to pass the following examination elements:

(a) Amateur Extra Class: Elements 1(C), 3, 4(A), and 4(B);

(b) Advanced Class: Elements 1(B), 3, and 4(A);

(c) General Class and Conditional Class: Elements 1(B) and 3;

(d) Technician Class: Elements 1(A) and 3;

(e) Novice Class: Elements 1(A) and 2.

§ 97.25 Examination credit.

(a) An applicant for a higher class of amateur operator license who holds a valid amateur operator license issued upon the basis of an examination by the Commission will be required to pass only those elements of the higher class examination that were not included in the examination for the amateur license held when such application was filed. However, credit will not be allowed for licenses issued on the basis of an examination given under the provisions of § 97.29(b).

(b) An applicant for an amateur operator license will be given credit for either telegraph code element 1(A) or 1(B) if within 5 years prior to the receipt of his application by the Commission he held a commercial radiotelegraph operator license or permit issued by the Federal Communications Commission. An applicant for an amateur extra class license will be given credit for the telegraph code element 1(C) if he holds a valid first class commercial radiotelegraph operator license or permit issued by the Federal Communications Commission or holds any commercial radiotelegraph operator license or permit issued by the Federal Communications Commission containing an aircraft radiotelegraph endorsement.

(c) An applicant for the Amateur Extra Class operator license will be given credit for examination elements 1(C), 4(A), and 4(B), if he so requests and submits evidence of having held a valid amateur radio station or operator license issued by any agency of the U.S. Government during or prior to April 1917, and qualifies for or currently holds a valid amateur operator license of the General or Advanced Class.

(d) An applicant for the amateur extra class operator license will be given credit for examination element 1(C) if he so requests and submits evidence of having held the amateur extra first class license, having continuously held its successor license. An applicant should present his proof in advance of the desired examination time to the Chief, Amateur and Citizens Division, Washington, D.C. 20554 and receive a letter of certification for presentation to the field office where the examination will be taken. No code credit will be given without the letter of certification.

(e) No examination credit, except as herein provided, shall be allowed on the basis of holding or having held any amateur or commercial operator license.

§ 97.27 Availability of Conditional Class license examinations.

The examinations for Conditional Class will be available only under one or more of the following conditions:

(a) If the applicant's actual residence and proposed amateur station location are more than 175 miles airline distance from the nearest location at which examinations are conducted by an authorized Commission employee or representative at intervals of not more than 6 months for amateur operator license.

(b) If the applicant is shown by physician's certificate to be unable to appear for examination because of protracted disability.

(c) If the applicant is shown by certificate of the commanding officer to be in the armed forces of the United States at an Army, Navy, Air Force, or Coast Guard station and, for that reason, to be unable to appear for examination at the time and place designated by the Commission.

(d) If the applicant demonstrates by sufficient evidence that his temporary residence is for a continuous period of at least 12 months outside the continental limits of the United States, its territories or possessions, irrespective of other provisions of this section.

§ 97.28 Mail examinations for disabled applicants for Amateur Extra and Advanced Class licenses.

(a) The Commission may permit the examination for an Amateur Extra or Advanced Class license to be administered by a volunteer examiner selected by the applicant when it is shown by a physician's certificate that the applicant is unable to appear for a Commission supervised examination because of protracted disability.

(b) The volunteer examiner for an Amateur Extra or Advanced Class license examination shall be at least 21 years of age and shall be the holder of a class of amateur operator license equal to or higher than the class of license for which the applicant is being examined. The written portion of the examination shall be obtained, supervised, and submitted in accordance with the procedures set forth in § 97.29(b).

§ 97.29 Manner of conducting examinations.

(a) Except as provided by § 97.28, the examination for Amateur Extra, Advanced and General Classes of amateur operator licenses will be conducted by an authorized Commission employee or representative at locations and at times specified by the Commission.

(b) Unless otherwise prescribed by the Commission, an examination for the Conditional, Technician, or Novice Class license will be conducted and supervised by a volunteer examiner selected by the applicant. A volunteer examiner shall be at least 21 years of age and shall be the holder of an Extra, Advanced, or General Class Amateur Radio operator license, or shall hold a Commercial radiotelegraph operator license issued by the Commission, or shall be employed in the service of the United States as the operator of a manually operated radiotelegraph station. The written portion of the examination shall be obtained, supervised, and submitted in accordance with the following procedure:

(1) Within 10 days after passing the required code test, an applicant shall submit an application (FCC Form 610), together with any filing fee prescribed, to the Commission's office at Gettysburg, Pennsylvania, 17325. The application shall include a written request from the volunteer examiner for the appropriate examination papers. The examiner's written request shall include (i) the names and permanent addresses of the examiner and the applicant, (ii) a description of the examiner's qualifications to administer the examination, (iii) the examiner's statement that the applicant has passed the code test for the class of license involved under his supervision within the 10 days prior to submission of the request, and (iv) the examiner's written signature. Examination papers will be forwarded only to the volunteer examiner.

NOTE: When the applicant is entitled to examination credit for the code test under one of the provisions of § 97.25, an application may be submitted without regard to the 10-day limitation. The examiner's request should then state that a code test was not administered for that reason. The applicant should furnish details as to the class, number, and expiration date of any Commercial radiotelegraph license involved.

(2) The volunteer examiner shall be responsible for the proper conduct and necessary supervision of the examination. Administration of the examination shall be in accordance with the instructions included with the examination papers and as prescribed in §§ 97.29 (c) and (d), 97.31, and 97.33.

(3) The examination papers, either completed or unopened in the event the examination is not taken, shall be returned by the volunteer examiner to the Commission's office at Gettysburg, Pa., no later than 30 days after the date the papers are mailed by the Commission (the date of mailing is normally stamped by the Commission on the outside of the examination envelope).

(c) The code test required of an applicant for amateur radio operator license, in accordance with the provisions of §§ 97.21 and 97.23 shall determine the applicant's ability to transmit by hand key (straight key or, if supplied by the applicant, any other type of hand operated key such as a semi-automatic or electronic key) and to receive by ear, in plain language, messages in the International Morse Code at not less than the prescribed speed, free from omission or other error for a continuous period of at least 1 minute during a test period of 5 minutes counting five characters to the word, each numeral or punctuation mark counting as two characters.

(d) All written portions of the examinations for amateur operator privileges shall be completed by the applicant in legible handwriting or hand printing, and diagrams shall be drawn by hand, by means of either pen and ink or pencil. Whenever the applicant's signature is required, his normal signature shall be used. Applicants unable to comply with these requirements, because of physical disability, may dictate their answers to the examination questions and the receiving code test and if unable to draw required diagrams, may dictate a detailed description essentially equivalent. If the examination or any part thereof is dictated, the examiner shall certify the nature of the applicant's disability and the name and address of the person(s) taking and transcribing the applicant's dictation.

§ 97.31 Grading of examinations.

(a) Code tests for sending and receiving are graded separately. Failure to pass the required code test for either sending or receiving will terminate the examination.

(b) Seventy-four percent (74%) is the passing grade for written examinations. For the purpose of grading, each element required in qualifying for a particular license will be considered as a separate examination. All written examinations will be graded only by Commission personnel.

§ 97.33 Eligibility for re-examination.

An applicant who fails examination for an amateur operator license may not take another examination for the same or a higher class amateur operator license within 30 days, except that this limitation shall not apply to an examination for an Advanced or General Class license following an examination conducted by a volunteer examiner for a Novice, Technician, or Conditional Class license.

§ 97.35 Additional examination for holders of operator licenses obtained by mail.

(a) A licensee who holds an amateur license which was obtained by a mail examination under the supervision of a volunteer examiner may be required to appear for a Commission supervised license examination at a location designated by the Commission. If the licensee fails to appear for this examination when directed to do so, or fails to pass such examination, the operator license involved shall be subject to cancellation. When a Novice, Technician, or Conditional Class license is cancelled under this provision, a new license will not be issued for the same class operator license as that cancelled.

(b) [Reserved]

(c) A holder of a Conditional Class license, obtained on the basis of an examination under the provisions of § 97.29(b), is not required to be re-examined when changing residence and station location to within a regular examination area, nor when a new examination location is established within 175 miles airline distance from such licensee's residence and station location.

Station Licenses

§ 97.37 General eligibility for station license.

An amateur radio station license will be issued only to a licensed amateur radio operator, except that a military recreation station license may also be issued to an individual not licensed as an amateur radio operator (other than an alien or a representative of an alien or of a foreign government), who is in charge of a proposed military recreation station not operated by the U.S. Government but which is to be located in approved public quarters.

§ 97.39 Eligibility of corporations or organizations to hold station license.

An amateur station license will not be issued to a school, company, corporation, association, or other organization, except that in the case of a bona fide amateur radio organization or society, a station license may be issued to a licensed amateur operator, other than the holder of a Novice Class license, as trustee for such society.

§ 97.40 Station license required.

(a) No transmitting station shall be operated in the amateur radio service without being licensed by the Federal Communications Commission.

(b) Every amateur radio operator must have a primary amateur radio station license.

(c) An amateur radio operator may be issued one or more additional station licenses, each for a different land location, except that repeater station, control station, and auxiliary link station licenses may also be issued to an amateur radio operator for land locations where another station license has been issued to the applicant.

(d) Any transmitter to be operated as part of a control link shall be licensed as a control station or as an auxiliary link station and may be combined with a primary, secondary, or club station license at the same location.

(e) A transmitter may only be operated as a repeater station under the authority of a repeater station license.

§ 97.41 Application for station license.

(a) Each application for a club or military recreation station license in the amateur radio service shall be made on the FCC Form 610–B. Each application for any other amateur radio station license shall be made on the FCC Form 610.

(b) Each application shall state whether the proposed station is a primary or additional station. If the latter, the application shall also state whether the proposed station is a secondary, control, auxiliary link, or repeater station.

(c) When an application(s) is made for a station having one or more associated stations, i.e., control station and/or auxiliary link station, a system network diagram shall also be submitted.

(d) Each application to license a remotely controlled amateur radio station, whether by wire or by radio control, shall be accompanied by a statement giving the address for each control point. The application shall include a functional block diagram and a technical explanation sufficient to describe the operation of the control link. Additionally, the following shall be provided:

(1) Description of the measures proposed for protection against access to the remote station by unauthorized persons.

(2) Description of the measures proposed for protection against unauthorized station operation, either through activation of the control link or otherwise.

(3) Description of the provisions for shutting down the station in case of control link malfunction.

(4) Description of the means to be provided for monitoring the transmitting frequencies.

(5) Photocopies of control station license(s) and auxiliary link station license(s), or the application(s) for same if such stations are proposed for the system network.

(e) Each application to license a control station or an auxiliary link station in the amateur radio service must be accompanied by the following information:

(1) The station transmitting band(s).

(2) Description of the means to be provided for monitoring the transmitting frequencies.

(3) The transmitter power input and justification that such power is in compliance with § 97.67(b).

(4) If remote control of an auxiliary link station is proposed, all of the information required by paragraph (d) of this section shall also be provided.

(f) [Reserved]

(g) One application and all papers incorporated therein and made a part thereof shall be submitted for each amateur station license. If the application is for station license only, it shall be filed directly with the Commission at its Gettysburg, Pa., office. If the application also contains application for any class of amateur operator license, it shall be filed in accordance with the provisions of § 97.11.

(h) Applicants proposing to construct a radio station on a site located on land under the jurisdiction of the U.S. Forest Service, U.S. Department of Agriculture, or the Bureau of Land Management, U.S. Department of the Interior, must supply the information and must follow the procedure prescribed by § 1.70 of this chapter.

(i) Each applicant in the Safety and Special Radio Services (1) for modification of a station license involving a site change or a substantial increase in tower height or (2) for a license for a new station must, before commencing construction, supply the environmental information, where required, and must follow the procedure prescribed by Subpart I of Part 1 of this chapter (§§ 1.301 through 1.1319) unless Commission action authorizing such construction would be a minor action with the meaning of Subpart I of Part 1.

§ 97.43 Location of station.

Every amateur station must have one land location, the address of which is designated on the station license. Every amateur radio station must have at least one control point. If the control point location is not the same as the station location, authority to operate the station by remote control is required.

§ 97.45 Limitations on antenna structures.

(a) Except as provided in paragraph (b) of this section, an antenna for a station in the Amateur Radio Service which exceeds the following height limitations may not be erected or used unless notice has been filed with both the FAA on FAA Form 7460-1 and with the Commission on Form 714 or on the license application form, and prior approval by the Commission has been obtained for:

(1) Any construction or alteration of more than 200 feet in height above ground level at its site (§ 17.7 (a) of this chapter).

(2) Any construction or alteration of greater height than an imaginary surface extending outward and upward at one of the following slopes (§ 17.7(b) of this chapter):

(i) 100 to 1 for a horizontal distance of 20,000 feet from the nearest point of the nearest runway of each airport with at least one runway more than 3,200 feet in length, excluding heliports and seaplane bases without specified boundaries, if that airport is either listed in the Airport Directory of the current Airman's Information Manual or is operated by a Federal military agency.

(ii) 50 to 1 for a horizontal distance of 10,000 feet from the nearest point of the nearest runway of each airport with its longest runway no more than 3,200 feet in length, excluding heliports and seaplane bases without specified boundaries, if that airport is either listed in the Airport Directory or is operated by a Federal military agency.

(iii) 25 to 1 for a horizontal distance of 5,000 feet from the nearest point of the nearest landing and takeoff area of each heliport listed in the Airport Directory or operated by a Federal military agency.

(3) Any construction or alteration on an airport listed in the Airport Directory of the Airman's Information Manual (§ 17.7(c) of this chapter).

(b) A notification to the Federal Aviation Administration is not required for any of the following construction or alteration:

(1) Any object that would be shielded by existing structures of a permanent and substantial character or by natural terrain or topographic features of equal or greater height, and would be located in the congested area of a city, town, or settlement where it is evident beyond all reasonable doubt that the structure so shielded will not adversely affect safety in air navigation. Applicants claiming such exemption shall submit a statement with their application to the Commission explaining the basis in detail for their finding (§ 17.14 (a) of this chapter).

(2) Any antenna structure of 20 feet or less in height except one that would increase the height of another antenna structure (§ 17.14(b) of this chapter).

(c) Further details as to whether an aeronautical study and/or obstruction marking and lighting may be required, and specifications for obstruction marking and lighting when required, may be obtained from Part 17 of this chapter, "Construction, Marking, and Lighting of Antenna Structures." Information regarding the inspection and maintenance of antenna structures requiring obstruction marking and lighting is also contained in Part 17 of this chapter.

§ 97.47 Renewal and/or modification of amateur station license.

(a) Application for renewal and/or modification of an individual station license shall be submitted on FCC Form 610, and application for renewal and/or modification of an amateur club or military recreation station shall be submitted on FCC Form 610-B. In every case the application shall be accompanied by the applicant's license or photocopy thereof. Applications for renewal of unexpired licenses must be made during the license term and should be filed not later than 60 days prior to the end of the license term. In any case in which the licensee has, in accordance with the provisions of this chapter, made timely and sufficient application for renewal of an unexpired license, no license with reference to any activity of a continuing nature shall expire until such application shall have been finally determined.

(b) If a license is allowed to expire, application for renewal may be made during a period of grace of 1 year after the expiration date. During this 1-year period of grace, an expired license is not valid. A license renewed during the grace period will be dated currently and will not be backdated to the date of expiration. An application for an individual station license shall be submitted on FCC Form 610. An application for an amateur club or military recreation station license shall be submitted on FCC Form 610-B. In every case the application shall be accompanied by the applicant's expired license or a photocopy thereof.

(c) When the name of a licensee is changed (without changes in the ownership, control, or corporate structure), or when the mailing address is changed (without changing the authorized location of the amateur radio station) a formal application for modification of license is not required. However, the licensee shall notify the Commission promptly of these changes. The notice, which may be in letter form, shall contain the name and address of the licensee as they appear in the Commission's records, the new name and/or address, as the case may be, and the call sign and the class of operator license. The notice shall be sent to Federal Communications Commission, Gettysburg, Pa., 17325, and a copy shall be maintained with the license of each station until a new license is issued.

(d) When an addition to the control point(s) authorized for a remotely controlled station is desired, an application for modification of the remotely controlled station license shall be submitted. Authorized control points may be deleted by letter notification to the Commission.

(e) Should the licensee desire to effect changes to his station which would significantly change the system network diagram or other technical and operational information on file with the Commission, revised showings for the proposed alterations shall be submitted for approval. An application for modification of the station license is not required.

§ 97.49 Commission modification of station license.

(a) Whenever the Commission shall determine that public interest, convenience, and necessity would be served, or any treaty ratified by the United States will be more fully complied with, by the modification of any radio station license either for a limited time, or for the duration of the term thereof, it shall issue an order for such licensee to show cause why such license should not be modified.

(b) Such order to show cause shall contain a statement of the grounds and reasons for such proposed modification, and shall specify wherein the said license is required to be modified. It shall require the licensee against whom it is directed to appear at a place and time therein named, in no event to be less than 30 days from the date of receipt of the order, to show cause why the proposed modification should not be made and the order of modification issued.

(c) If the licensee against whom the order to show cause is directed does not appear at the time and place provided in said order, a final order of modification shall issue forthwith.

Call Signs

§ 97.51 Assignment of call signs.

(a) The call signs of amateur stations will be assigned systematically by the Commission with the following exceptions:

(1) A specific unassigned call sign may be reassigned to the most recent holder thereof;

(2) A specific unassigned call sign may be assigned to a previous holder if not under license during the past 5 years;

(3) A specific unassigned call sign may be assigned to an amateur organization in memoriam to a deceased member and former holder thereof;

(4) A specific call sign may be temporarily assigned to a station connected with an event, or events, of general public interest;

(5) One unassigned two-letter call sign (a call sign having two letters following the numeral) may be assigned to a previous holder of a two-letter call sign, the prefix of which consisted of not more than a single letter. Additionally, a two-letter call sign may be assigned to an Amateur Extra Class licensee who submits evidence that he held any amateur radio operator or station license, issued by any agency of the U.S. Government or by any foreign government, 25 years or more prior to the receipt date of an application for such assignment. Applicants for two-letter call signs are not permitted to select a specific assignment except in accordance with subparagraphs (1) and (2) of this paragraph.

(b) An amateur call sign will consist of a sequence of one or two letters, a numeral designating the call sign area, and two or three letters. The call sign areas are as follows:

1. Maine, New Hampshire, Vermont, Massachusetts, Rhode Island, Connecticut.
2. New York, New Jersey.
3. Pennsylvania, Delaware, Maryland, District of Columbia.
4. Virginia, North and South Carolina, Georgia, Florida, Alabama, Tennessee, Kentucky, Puerto Rico and Virgin Islands.
5. Mississippi, Louisiana, Arkansas, Oklahoma, Texas, New Mexico.
6. California, Hawaii and Pacific possessions except those included in area 7.
7. Oregon, Washington, Idaho, Montana, Wyoming, Arizona, Nevada, Utah, Alaska and adjacent islands.
8. Michigan, Ohio, West Virginia.
9. Wisconsin, Illinois, Indiana.
10. Colorado, Nebraska, North and South Dakota, Kansas, Minnesota, Iowa, Missouri.

§ 97.53 Policies and procedures applicable to assignment of call signs.

(a) The following are regarded as preferred call signs:

(1) Two-letter call signs—call signs with a single letter prefix (two-letter prefix in Alaska, Hawaii, and in the U.S. possessions) and a two-letter suffix; e.g. W6AB (KH6AB).

(2) Three-letter call signs—call signs with a single letter prefix and a three-letter suffix; e.g. W6ABC.

(b) An eligible licensee will be permitted to hold only one two-letter call sign. However, a licensee who, by reason of former rule provisions, presently holds more than one such call sign may continue to hold those call signs in the same call sign areas.

(c) Subject to availability, two-letter call signs beginning with the letter "W" will normally be assigned in each call sign area to eligible licensees.

(d) An eligible licensee who holds one or more three-letter call signs must relinquish one of those call signs in order to be assigned a two-letter call sign.

(e) New additional stations will not be assigned a preferred call sign.

(f) An additional station which is presently assigned a preferred call sign will be issued a nonpreferred call sign upon modification of license to show a station location in a different call sign area.

(g) Subject to availability, a basic station will be issued the same type of call sign as the one relinquished upon modification of license to show a station location in a different call sign area.

(1) Licensees will not be assigned specific call signs or their choice of counterpart call signs (call signs with identical suffix letters) under this provision.

(2) When a two-letter call sign is not available in the new call sign area, an eligible licensee may be assigned an available unspecified three-letter call sign.

(h) Call signs which have been unassigned for more than one year are normally available for reassignment.

Duplicate Licenses and License Term

§ 97.57 Duplicate license.

Any licensee requesting a duplicate license to replace an original which has been lost, mutilated, or destroyed, shall submit a statement setting forth the facts regarding the manner in which the original license was lost, mutilated, or destroyed. If, subsequent to receipt by the licensee of the duplicate license, the original license is found, either the duplicate or the original license shall be returned immediately to the Commission.

§ 97.59 License term.

(a) Amateur operator licenses are normally valid for a period of 5 years from the date of issuance of a new or renewed license, except the Novice Class which is normally valid for a period of 2 years from the date of issuance.

(b) The license for an amateur station is normally valid for a period of 5 years from the date of issuance of a new or renewed license, except that an amateur station license issued to the holder of a Novice Class amateur operator license is normally valid for a period of 2 years from the date of issuance. All amateur station licenses, regardless of when issued, will expire on the same date as the licensee's amateur operator license.

(c) A duplicate license or a modified license which is not being renewed shall bear the same expiration date as the license for which it is a modification or duplicate.

SUBPART C—TECHNICAL STANDARDS

§ 97.61 Authorized frequencies and emissions.

(a) Following are the frequency bands and associated emissions available to amateur radio stations, other than repeater stations, subject to the limitations stated in paragraph (b) of this section, §§ 97.65, 97.109, and 97.110.

[See next column]

(b) Limitations:

(1) The use of frequencies in this band is on a shared basis with the LORAN–A radionavigation system and is subject to cancellation or revision, in whole or in part, by order of the Commission, without hearing, whenever

Authorized frequencies and emissions.

Frequency band	Emissions	Limitation (see paragraph (b))
kHz		
1800-2000	A1, A3	1,2
3500-4000	A1	
3500-3775	F1	
3775-3890	A5, F5	
3775-4000	A3, F3	4
4383.8	A3J/A3A	13
7000-7300	A1	3,4
7000-7150	F1	3,4
7075-7100	A3, F3	11
7150-7225	A5, F5	3,4
7150-7300	A3, F3	3,4
14000-14350	A1	
14000-14200	F1	
14200-14275	A5, F5	
14200-14350	A3, F3	
MHz		
21.000-21.450	A1	
21.000-21.250	F1	
21.250-21.350	A5, F5	
21.250-21.450	A3, F3	
28.000-29.700	A1	
28.000-28.500	F1	
28.500-29.700	A3, F3, A5, F5	
50.0-54.0	A1	
50.1-54.0	A2, A3, A4, A5, F1, F3, F5	
51.0-54.0	AØ	
144-148	A1	
144.1-148.0	AØ, A2, A3, A4, A5, FØ, F1, F2, F3, F5	
220-255	AØ, A1, A2, A3, A4, A5, FØ, F1, F2, F3, F4, F5.	5,6
420-450	AØ, A1, A2, A3, A4, A5, FØ, F1, F2, F3, F4, F5.	5,7
1215-1300	AØ, A1, A2, A3, A4, A5, FØ, F1, F2, F3, F4, F5.	5
2300-2450	AØ, A1, A2, A3, A4, A5, FØ, F1, F2, F3, F4, F5, P.	5,8
3300-3500	AØ, A1, A2, A3, A4, A5, FØ, F1, F2, F3, F4, F5, P.	5,12
5650-5925	AØ, A1, A2, A3, A4, A5, FØ, F1, F2, F3, F4, F5, P.	5,9
GHZ		
10.000-10.500	AØ, A1, A2, A3, A4, A5, FØ, F1, F2, F3, F4, F5.	5
24.000-24.250	AØ, A1, A2, A3, A4, A5, FØ, F1, F2, F3, F4, F5, P.	5,10
48.000-50.000	AØ, A1, A2, A3, A4, A5, FØ, F1, F2, F3, F4, F5, P.	
71.000-84.000	AØ, A1, A2, A3, A4, A5, FØ, F1, F2, F3, F4, F5, P.	
152.00-170.00	AØ, A1, A2, A3, A4, A5, FØ, F1, F2, F3, F4, F5, P.	
200.00-220.00	AØ, A1, A2, A3, A4, A5, FØ, F1, F2, F3, F4, F5, P.	
240.00-250.00	AØ, A1, A2, A3, A4, A5, FØ, F1, F2, F3, F4, F5, P.	
Above 275.00	AØ, A1, A2, A3, A4, A5, FØ, F1, F2, F3, F4, F5, P.	

the Commission shall determine such action is necessary in view of the priority of the LORAN–A radionavigation system. The use of these frequencies by amateur stations shall not cause harmful interference to LORAN–A system. If an amateur station causes such interference, operation on the frequencies involved must cease if so directed by the Commission.

(2) *[See next page]*

(3) Where, in adjacent regions or subregions, a band of frequencies is allocated to different services of the same category, the basic principle is the equality of right to operate. Accordingly, the stations of each service in one region or subregion must operate so as not to cause harmful interference to services in the other regions or subregions (No. 117, the Radio Regulations, Geneva, 1959).

(4) 3900–4000 kHz and 7100–7300 kHz are not available in the following U.S. possessions: Baker, Canton, Enderbury, Guam, Howland, Jarvis, Palmyra, American Samoa, and Wake Islands.

(5) Amateur stations shall not cause interference to the Government radiolocation service.

(6) Not available in those portions of Texas and New Mexico bounded by latitude 33°24′ N., and 31°53′

(2) Operation shall be limited to:

Area	Maximum DC plate input power in watts							
	1800–1825 kHz	1825–1850 kHz	1850–1875 kHz	1875–1900 kHz	1900–1925 kHz	1925–1950 kHz	1950–1975 kHz	1975–2000 kHz
	Day/Night	Day/Night	Day/Night	Day/Night	Day/Night	Day/Night	Day/Night	Day/Night
Alabama	500/100	100/25	0	0	0	0	100/25	500/100
Alaska	1000/200	500/100	500/100	100/25	0	0	0	0
Arizona	1000/200	500/100	500/100	0	0	0	0	0
Arkansas	1000/200	500/100	100/25	0	0	100/25	100/25	500/100
California	1000/200	500/100	500/100	100/25	0	0	0	0
Colorado	1000/200	500/100	200/50	0	0	0	0	200/50
Connecticut	500/100	100/25	0	0	0	0	0	0
Delaware	500/100	100/25	0	0	0	0	0	100/25
District of Columbia	500/100	100/25	0	0	0	0	0	100/25
Florida	500/100	100/25	0	0	0	0	100/25	500/100
Georgia	500/100	100/25	0	0	0	0	0	200/50
Hawaii	0	0	0	0	200/50	100/25	100/25	500/100
Idaho	1000/200	500/100	500/100	100/25	100/25	100/25	100/25	500/100
Illinois	1000/200	500/100	100/25	0	0	0	0	200/50
Indiana	1000/200	500/100	100/25	0	0	0	0	200/50
Iowa	1000/200	500/100	200/50	0	0	100/25	100/25	500/100
Kansas	1000/200	500/100	100/25	0	0	100/25	100/25	500/100
Kentucky	1000/200	500/100	100/25	0	0	0	0	200/50
Louisiana	500/100	100/25	0	0	0	0	100/25	500/100
Maine	500/100	100/25	0	0	0	0	0	0
Maryland	500/100	100/25	0	0	0	0	0	100/25
Massachusetts	500/100	100/25	0	0	0	0	0	0
Michigan	1000/200	500/100	100/25	0	0	0	0	100/25
Minnesota	1000/200	500/100	500/100	100/25	100/25	100/25	100/25	500/100
Mississippi	500/100	100/25	0	0	0	0	100/25	500/100
Missouri	1000/200	500/100	100/25	0	0	100/25	100/25	500/100
Montana	1000/200	500/100	500/100	100/25	100/25	100/25	100/25	500/100
Nebraska	1000/200	500/100	200/50	0	0	100/25	100/25	500/100
Nevada	1000/200	500/100	500/100	100/25	0	0	0	0
New Hampshire	500/100	100/25	0	0	0	0	0	0
New Jersey	500/100	100/25	0	0	0	0	0	0
New Mexico	1000/200	500/100	100/25	0	0	100/25	500/100	1000/200
New York	500/100	100/25	0	0	0	0	0	0
North Carolina	500/100	100/25	0	0	0	0	0	100/25
North Dakota	1000/200	500/100	500/100	100/25	100/25	100/25	100/25	500/100
Ohio	1000/200	500/100	100/25	0	0	0	0	100/25
Oklahoma	1000/200	500/100	100/25	0	0	100/25	100/25	500/100
Oregon	1000/200	500/100	500/100	100/25	0	0	0	0
Pennsylvania	500/100	100/25	0	0	0	0	0	0
Rhode Island	500/100	100/25	0	0	0	0	0	0
South Carolina	500/100	100/25	0	0	0	0	0	200/50
South Dakota	1000/200	500/100	500/100	100/25	100/25	100/25	100/25	500/100
Tennessee	1000/200	500/100	100/25	0	0	0	0	200/50
Texas	500/100	100/25	0	0	0	0	0	200/50
Utah	1000/200	500/100	500/100	100/25	100/25	0	0	100/25
Vermont	500/100	100/25	0	0	0	0	0	0
Virginia	500/100	100/25	0	0	0	0	0	100/25
Washington	1000/200	500/100	500/100	100/25	0	0	0	0
West Virginia	1000/200	500/100	100/25	0	0	0	0	100/25
Wisconsin	1000/200	500/100	200/50	0	0	0	0	200/50
Wyoming	1000/200	500/100	500/100	100/25	100/25	0	0	200/50
Puerto Rico	500/100	100/25	0	0	0	0	0	200/50
Virgin Islands	500/100	100/25	0	0	0	0	0	200/50
Swan Island	500/100	100/25	0	0	0	0	100/25	500/100
Serrana Bank	500/100	100/25	0	0	0	0	100/25	500/100
Roncador Key	500/100	100/25	0	0	0	0	100/25	500/100
Navassa Island	500/100	100/25	0	0	0	0	0	200/50
Baker, Canton, Enderbury, Howland	100/25	0	0	100/25	100/25	0	0	100/25
Guam, Johnston, Midway	0	0	0	0	100/25	0	0	100/25
American Samoa	200/50	0	0	200/50	200/50	0	0	200/50
Wake	100/25	0	0	100/25	0	0	0	0
Palmyra, Jarvis	0	0	0	0	200/50	0	0	200/50

N., and longitude 105°40′ W. and 106°40′ W. between the hours 0500 and 1800 local time, Monday through Friday, except to stations authorized to operate in an organized civil defense network when civil defense emergencies exist or when arrangements have been made with the Commission Engineer in Charge at Dallas, Tex., and the Area Frequency Coordinator at White Sands, N. Mex., for drills at specific dates and times.

(7) In the following areas the d.c. plate input power to the final transmitter stage shall not exceed 50 watts, except when authorized by the appropriate Commission Engineer in Charge and the appropriate Military Area Frequency Coordinator.

(i) Those portions of Texas and New Mexico bounded by latitude 33°24′ N., 31°53′ N., and longitude 105°40′ W. and 106°40′ W.

(ii) The State of Florida, including the Key West area and the areas enclosed within circles of 200-mile radius centered at 28°21′ N., 80°43′ W. and 30°30′ N., 86°30′ W.

(iii) The State of Arizona.

(iv) Those portions of California and Nevada south of latitude 37°10′ N. and the area within a 200-mile radius of 34°09′ N., 119°11′ W.

(8) No protection in the band 2400–2500 MHz is afforded from interference due to the operation of industrial, scientific, and medical devices on 2450 MHz.

(9) No protection in the band 5725–5875 MHz is afforded from interference due to the operation of industrial, scientific and medical devices on 5800 MHz.

(10) No protection in the band 24.00–24.25 GHz is afforded from interference due to the operation of industrial, scientific and medical devices on 24.125 GHz.

(11) The use of A3 and F3 in this band is limited to amateur radio stations located outside Region 2.

(12) Amateur stations shall not cause interference to the Fixed-Satellite Service operating in the band 3400–3500 MHz.

(13) The frequency 4383.8 kHz, maximum power 150 watts, may be used by any station authorized under this part to communicate with any other station authorized in the State of Alaska for emergency communications. No airborne operations will be permitted on this frequency. Additionally, all stations operating on this frequency must be located in or within 50 nautical miles of the State of Alaska.

(c) The following transmitting frequency bands and the associated emission authorized in paragraph (a) of this section are available for repeater stations, including both input (receiving) and output (transmitting):

Frequency Band (MHz)

52.0–54.0
146.0–148.0
222.0–225.0
442.0–450.0

any amateur frequency above 1215 MHz.

The frequency band 29.5–29.7 MHz may be authorized upon a special showing of need for repeater station operation in this band for intracommunity amateur radio communications.

§ 97.63 Individual frequency not specified.

Transmissions by an amateur station may be on any frequency within any authorized amateur band. Sideband frequencies resulting from keying or modulating a carrier wave shall be confined within the authorized amateur band.

§ 97.65 Emission limitations.

(a) Type AØ emission, where not specifically designated in the bands listed in § 97.61, may be used for short periods of time when required for authorized remote control purposes or for experimental purposes. However, these limitations do not apply where type AØ emission is specifically designated.

(b) Whenever code practice, in accordance with § 97.91(d), is conducted in bands authorized for A3 emission, tone modulation of the radiotelephone transmitter may be utilized when interspersed with appropriate voice instructions.

(c) On frequencies below 29.0 MHz and between 50.1 and 52.5 MHz, the bandwidth of an F3 emission (frequency or phase modulation) shall not exceed that of an A3 emission having the same audio characteristics; and the purity and stability of emissions shall comply with the requirements of § 97.73.

(d) On frequencies below 50 MHz, the bandwidth of A5 and F5 emissions shall not exceed that of an A3 single sideband emission.

(e) On frequencies between 50 MHz and 225 MHz, single sideband or double sideband A5 emission may be used and the bandwidth shall not exceed that of an A3 single sideband or double sideband signal respectively. The bandwidth of F5 emission shall not exceed that of an A3 single sideband emission.

(f) Below 225 MHz, A3 and A5 emissions may be used simultaneously on the same carrier frequency provided the total bandwidth does not exceed that of an A3 double sideband emission.

§ 97.67 Maximum authorized power.

(a) Except for power restrictions as set forth in § 97.61, each amateur transmitter may be operated with a power input not exceeding 1 kilowatt to the plate circuit of the final amplifier stage of an amplifier-oscillator transmitter or to the plate circuit of an oscillator transmitter. An amateur transmitter operating with a power input exceeding 900 watts to the plate circuit shall provide means for accurately measuring the plate power input to the vacuum tube or tubes supplying power to the antenna.

(b) Notwithstanding the provisions of paragraph (a) of this section, amateur stations shall use the minimum amount of transmitter power necessary to carry out the desired communications.

(c) Within the limitations of paragraphs (a) and (b) of this section, the effective radiated power of a repeater station shall not exceed that specified for the antenna height above average terrain in the following table:

Antenna height above average terrain	Maximum effective radiated power for frequency bands above:			
	52 MHz	146 MHz	442 MHz	1215 MHz
Below 50 feet	100 watts	800 watts	Paragraphs (a) and (b).	
50 to 99 feet	100 watts	400 watts	do	
100 to 499 feet	50 watts	400 watts	800 watts	Paragraphs (a) and (b).
500 to 999 feet	25 watts	200 watts	800 watts	Do.
Above 1,000 feet	25 watts	100 watts	400 watts	Do.

§ 97.69 Radio teleprinter transmissions.

The following special conditions shall be observed during the transmission of radio teleprinter signals on authorized frequencies by amateur stations:

(a) A single channel five-unit (start-stop) teleprinter code shall be used which shall correspond to the International Telegraphic Alphabet No. 2 with respect to all letters and numerals (including the slant sign or fraction bar) but special signals may be employed for the remote control of receiving printers, or for other purposes, in "figures" positions not utilized for numerals. In general, this code shall conform as nearly as possible to the teleprinter code or codes in common commercial usage in the United States.

(b) The normal transmitting speed of the radio teleprinter signal keying equipment shall be adjusted as closely as possible to one of the standard teleprinter speeds, namely, 60 (45 bauds), 67 (50 bauds), 75 (56.25 bauds) or 100 (75 bauds) words per minute, and in any event, within the range of ±5 words per minute of the selected standard speed.

(c) When frequency shift keying (type F1 emission) is utilized, the deviation in frequency from the mark signal to space signal, or from the space signal to the mark signal, shall be less than 900 hertz.

(d) When audio frequency shift keying (type A2 or type F2 emission) is utilized, the highest fundamental modulating audio frequency shall not exceed 3000 hertz, and the difference between the modulating audio frequency for the mark signal and that for the space signal shall be less than 900 hertz.

§ 97.71 Transmitter power supply.

The licensee of an amateur station using frequencies below 144 megahertz shall use adequately filtered direct-current plate power supply for the transmitting equipment to minimize modulation from this source.

§ 97.73 Purity and stability of emissions.

Spurious radiation from an amateur station being operated with a carrier frequency below 144 megahertz shall be reduced or eliminated in accordance with good engineering practice. This spurious radiation shall not be of sufficient intensity to cause interference in receiving equipment of good engineering design including adequate selectivity characteristics, which is tuned to a frequency or frequencies outside the frequency band of emission normally required for the type of emission being employed by the amateur station. In the case of A3 emission, the amateur transmitter shall not be modulated to the extent that interfering spurious radiation occurs, and in no case shall the emitted carrier wave be amplitude-modulated in excess of 100 percent. Means shall be employed to insure that the transmitter is not modulated in excess of its modulation capability for proper technical operation. For the purposes of this section a spurious radiation is any radiation from a transmitter which is outside the frequency band of emission normal for the type of transmission employed, including any component whose frequency is an integral multiple or submultiple of the carrier frequency (harmonics and subharmonics), spurious modulation products, key clicks, and other transient effects, and parasitic oscillations. When using amplitude modulation on frequencies below 144 megahertz, simultaneous frequency modulation is not permitted and when using frequency modulation on frequencies below 144 megahertz simultaneous amplitude modulation is not permitted. The frequency of the emitted carrier wave shall be as constant as the state of the art permits.

§ 97.75 Frequency measurement and regular check.

The licensee of an amateur station shall provide for measurement of the emitted carrier frequency or frequencies and shall establish procedure for making such measurement regularly. The measurement of the emitted carrier frequency or frequencies shall be made by means independent of the means used to control the radio frequency or frequencies generated by the transmitting apparatus and shall be of sufficient accuracy to assure operation within the amateur frequency band used.

SUBPART D—OPERATING REQUIREMENTS AND PROCEDURES

GENERAL

§ 97.77 Practice to be observed by all licensees.

In all respects not specifically covered by these regulations each amateur station shall be operated in accordance with good engineering and good amateur practice.

§ 97.79 Control operator requirements.

(a) The licensee of an amateur station shall be responsible for its proper operation.

(b) Every station when in operation shall have a control operator at an authorized control point. The control operator may be the station licensee or another amateur radio operator designated by the licensee. Each control operator shall also be responsible for the proper operation of the station.

(c) An amateur station may only be operated in the manner and to the extent permitted by the operator privileges authorized for the class of license held by the control operator, but may exceed those of the station licensee provided proper station identification procedures are performed.

(d) The licensee of an amateur radio station may permit any third party to participate in amateur radio communication from his station, provided that a control operator is present and continuously monitors and supervises the radio communication to insure compliance with the rules.

§ 97.81 Authorized apparatus.

An amateur station license authorizes the use under control of the licensee of all transmitting apparatus at the fixed location specified in the station license which is operated on any frequency, or frequencies allocated to the amateur service, and in addition authorizes the use, under control of the licensee, of portable and mobile transmitting apparatus operated at other locations.

§ 97.83 Availability of operator license.

The original operator license of each operator shall be kept in the personal possession of the operator while operating an amateur station. When operating an amateur station at a fixed location, however, the license may be posted in a conspicuous place in the room occupied by the operator. The license shall be available for inspection by any authorized Government official whenever the operator is operating an amateur station and at other times upon request made by an authorized representative of the Commission, except when such license has been filed with application for modification or renewal thereof, or has been mutilated, lost or destroyed, and request has been made for a duplicate license in accordance with § 97.57. No recognition shall be accorded to any photocopy of an operator license; however, nothing in this section shall be construed to prohibit the photocopying for other purposes of any amateur radio operator license.

§ 97.85 Availability of station license.

The original license of each amateur station or a photocopy thereof shall be posted in a conspicuous place in the room occupied by the licensed operator while the station is being operated at a fixed location or shall be kept in his personal possession. When the station is operated at other than a fixed location, the original station license or a photocopy thereof shall be kept in the personal possession of the station licensee (or a licensed representative) who shall be present at the station while it is being operated as a portable or mobile station. The original station license shall be available for inspection by any authorized Government official at all times while the station is being operated and at other times upon request made by an authorized representative of the Commission, except when such license has been filed with application for modification or renewal thereof, or has been mutilated, lost, or destroyed, and request has been made for a duplicate license in accordance with § 97.57.

§ 97.87 Station identification.

(a) An amateur station shall be identified by the transmission of its call sign at the beginning and end of each single transmission or exchange of transmissions and at intervals not to exceed 10 minutes during any single transmission or exchange of transmissions of more than 10 minutes duration. Additionally, at the end of an exchange of telegraphy (other than teleprinter) or telephony transmissions between amateur stations, the call sign (or the generally accepted network identifier) shall be given for the station, or for at least one of the group of stations, with which communication was established.

(b) When an amateur station is operated as a portable or mobile station, the operator shall give the following additional identification at the end of each single transmission or exchange of transmissions:

(1) When identifying by telegraphy, immediately after the call sign, transmit the fraction-bar $\overline{DN}$ followed by the number of the call sign area in which the station is being operated.

(2) When identifying by telephony, immediately after the call sign, transmit the word "portable" or "mobile", as appropriate, followed by the number of the call sign area in which the station is being operated.

(c) When an amateur station is operated outside of the 10 call sign areas prescribed in § 97.51(b) and outside of the jurisdiction of a foreign government, the operator shall give the following additional identification at the end of each single transmission or exchange of transmissions:

(1) When identifying by telegraphy, immediately after the call sign, transmit the fraction-bar $\overline{DN}$ fol-

lowed by the designator R 1, R 2, or R 3, to show the region (as defined by the International Radio Regulations, Geneva, 1959) in which the station is being operated.

(2) When identifying by telephone, immediately after the call sign, transmit the word "mobile" followed by the designator Region 1, Region 2, or Region 3, to show the region (as defined by the International Radio Regulations, Geneva, 1959) in which the station is being operated.

(d) Under conditions when the control operator is other than the station licensee, the station identification shall be the assigned call sign for that station. However, when a station is operated within the privileges of the operator's class of license but which exceeds those of the station licensee, station identification shall be made by following the station call sign with the operator's primary station call sign (i.e. WN4XYZ/W4XX).

(e) A repeater station shall be identified by radiotelephony or by radio telegraphy when in service at intervals not to exceed 5 minutes at a level of modulation sufficient to be intelligible through the repeated transmission.

(f) A control station must be identified by its assigned station call sign unless its emissions contain the call sign identification of the remotely controlled station.

(g) An auxiliary link station must be identified by its assigned station call sign unless its emissions contain the call sign of its associated station.

(h) The identification required by paragraphs (a), (b), (c), (d), (e), (f), and (g) of this section shall be given on each frequency being utilized for transmission and shall be transmitted either by telegraphy using the international Morse code, or by telephony, using the English language. If by an automatic device only used for identification by telegraphy, the code speed shall not exceed 20 words per minute. The use of a national or internationally recognized standard phonetic alphabet as an aid for correct telephone identification is encouraged.

§ 97.88 Operation of a remotely controlled station.

An amateur radio station may be remotely controlled only from an authorized control point, and only where there is compliance with the following:

(a) The license for the remotely controlled station must list the authorized remote control point(s). A photocopy of the remotely controlled station license must be posted in a conspicuous place at the authorized control point(s), and at the remotely controlled transmitter location. A copy of the system network diagram on file with the Commission must be retained at each control point. The transmitting antenna, transmission line, or mast, as appropriate, associated with the remotely controlled transmitter must bear a durable tag marked with the station call sign, the name of the station licensee and other information so that the control operator can readily be contacted by Commission personnel.

(b) The control link equipment and the remotely controlled station must be accessible only to persons authorized by the licensee. Protection against both inadvertent and unauthorized deliberate emissions must be provided. In the event unauthorized emissions occur, the station operation must be suspended until such time as adequate protection is incorporated, or until there is reasonable assurance that unauthorized emissions will not recur.

(c) A control operator designated by the licensee must be on duty at an authorized control point while the station is being remotely controlled. Immediately prior to, and during the periods the remotely controlled station is in operation, the frequencies used for emission by the remotely controlled transmitter must be continuously monitored by the control operator. The control operator must terminate transmission upon any deviation from the rules.

(d) Provisions must be incorporated to limit transmission to a period of no more than 3 minutes in the event of malfunction in the control link.

(e) A remotely controlled station may not be operated at any location other than that specified on the license without prior approval of the Commission except in emergencies involving the immediate safety of life or protection of property.

(f) A repeater station may be operated by radio remote control only where the control link utilizes frequencies other than the repeater station receiving frequencies.

§ 97.89 Points of communications.

(a) Amateur stations may communicate with:

(1) Other amateur stations, excepting those prohibited by Appendix 2.

(2) Stations in other services licensed by the Commission and with U.S. Government stations for civil defense purposes in accordance with Subpart F of this part, in emergencies and, on a temporary basis, for test purposes.

(3) Any station which is authorized by the Commission to communicate with amateur stations.

(b) Amateur stations may be used for transmitting signals, or communications, or energy, to receiving apparatus for the measurement of emissions, temporary observation of transmission phenomena, radio control of remote objects, and similar experimental purposes and for the purposes set forth in § 97.91.

(c) Notwithstanding the provisions of paragraph (a), no more than two repeater stations may operate in tandem, i.e., one repeating the transmissions of the other, excepting emergency operations provided for in § 97.107 or brief periods to conduct emergency preparedness tests.

(d) Control stations and auxiliary link stations may not be used to communicate with any other station than those shown in the system network diagram.

§ 97.91 One-way communications.

In addition to the experimental one-way transmission permitted by § 97.89, the following kinds of one-way communications, addressed to amateur stations, are authorized and will not be construed as broadcasting: (a) Emergency communications, including bona-fide emergency drill practice transmissions; (b) Information bulletins consisting solely of subject matter having direct interest to the amateur radio service as such; (c) Round-table discussions or net-type operations where more than two amateur stations are in communication, each station taking a turn at transmitting to other station(s) of the group; and (d) Code practice transmissions intended for persons learning or improving proficiency in the International Morse Code.

§ 97.93 Modulation of carrier.

Except for brief tests or adjustments, an amateur radiotelephone station shall not emit a carrier wave on frequencies below 51 megahertz unless modulated for the purpose of communication. Single audiofrequency tones may be transmitted for test purposes of short duration for the development and perfection of amateur radio telephone equipment.

STATION OPERATION AWAY FROM AUTHORIZED LOCATION

§ 97.95 Operation away from the authorized permanent station location.

(a) Operation within the United States, its territories, or possessions is permitted as follows:

(1) When there is no change in the authorized land station location, an amateur radio station other than a military recreation or an auxiliary link station may be operated under its station license anywhere in the United States, its territories or possessions as a portable or mobile operation, subject to § 97.61.

(2) When the authorized permanent station location is changed, formal application (FCC Form 610 for an individual station license and FCC Form 610–B for an amateur club or military recreation station license) must be submitted to the Commission prior to any operation and within 4 months of the move for the purpose of modifying the station license to show the new permanent station location. Operation at the new location is permitted under the license for the former station from the date the modification application is mailed until advised of Commission action on that application.

(3) For operations under subparagraphs (1) and (2) of this paragraph, advance notice, as required by § 97.97, must be given to the Engineer in Charge of each radio district in which operation is intended and the portable identification procedures specified in § 97.87 must be used.

(b) When outside the continental limits of the United States, its territories, or possessions, an amateur radio station may be operated as portable or mobile only under the following conditions:

(1) Operation may not be conducted within the jurisdiction of a foreign government except pursuant to, and in accordance with express authority granted to the licensee by such foreign government. When a foreign government permits Commission licensees to operate within its territory, the amateur frequency bands which may be used shall be as prescribed or limited by that government. (See Appendix 4 of this Part for the text of treaties or agreements between the United States and foreign governments relative to reciprocal amateur radio operation.)

(2) When outside the jurisdiction of a foreign government, operation may be conducted within Region 2 on any amateur frequency band between 7.0 MHz and 148 MHz, inclusive; and when not within Region 2, operation may be conducted only in the amateur bands 7.0–7.1 MHz, 14.00–14.35 MHz, 21.00–21.45 MHz, and 28.0–29.7 MHz.

NOTE: Region 2 is defined as follows: On the east, a line (B) extending from the North Pole along meridian 10° west of Greenwich to its intersection with parallel 72° north; thence by Great Circle Arc to the intersection of meridian 50° west and parallel 40° north; thence by Great Circle Arc to the intersection of meridian 20° west and parallel 10° south; thence along meridian 20° west to the South Pole. On the west, a line (C) extending from the North Pole by Great Circle Arc to the intersection of parallel 65°30′ north with the international boundary in Bering Strait; thence by Great Circle Arc to the intersection of meridian 165° east of Greenwich and parallel 50° north; thence by Great Circle Arc to the intersection of meridian 170° west and parallel 10° north; thence along parallel 10° north to its intersection with meridian 120° west; thence along meridian 120° west to the South Pole.

(3) Notice of such operation, in accordance with the provisions of § 97.97, shall be given to the Engineer in Charge of the district having jurisdiction of the authorized fixed transmitter location.

§ 97.97 Notice of operation away from authorized location.

Whenever an amateur station is, or is likely to be, in portable operation at a single location for a period exceeding 15 days, the licensee shall give advanced written notice of such operation to the Commission's office specified in § 97.95. A new notice is required whenever there is any change in the particulars of a previous notice or whenever operation away from the authorized station continues for a period in excess of 1 year. The notice required by this section shall contain the following information:

(a) Name of licensee.

(b) Station call sign.

(c) Authorized station location shown on station license.

(d) Specific geographical location of station when in portable operation.

(e) Dates of the beginning and end of the portable operation.

(f) Address at which, or through which, the licensee can be readily reached.

SPECIAL PROVISIONS

§ 97.99 Stations used only for radio control of remote model crafts and vehicles.

An amateur transmitter when used for the purpose of transmitting radio signals intended only for the control of a remote model craft or vehicle and having mean output power not exceeding one watt may be operated under the special provisions of this section provided an executed Transmitter Identification Card (FCC Form 452–C) or a plate made of a durable substance indicating the station call sign and licensee's name and address is affixed to the transmitter.

(a) Station identification is not required for transmissions directed only to a remote model craft or vehicle.

(b) Transmissions containing only control signals directed only to a remote model craft or vehicle are not considered to be codes or ciphers in the context of the meaning of § 97.117.

(c) Notice of operation away from authorized location is not required where the portable or mobile operation consists entirely of transmissions directed only to a remote model craft or vehicle.

(d) Station logs need not indicate the times of commencing and terminating each transmission or series of transmissions.

§ 97.101 Mobile stations aboard ships or aircraft.

In addition to complying with all other applicable rules, an amateur mobile station operated on board a ship or aircraft must comply with all of the following special conditions: (a) The installation and operation of the amateur mobile station shall be approved by the master of the ship or captain of the aircraft; (b) The amateur mobile station shall be separate from and independent of all other radio equipment, if any, installed on board the same ship or aircraft; (c) The electrical installation of the amateur mobile station shall be in accord with the rules applicable to ships or aircraft as promulgated by the appropriate government agency; (d) The operation of the amateur mobile station shall not interfere with the efficient operation of any other radio equipment installed on board the same ship or aircraft; and (e) The amateur mobile station and its associated equipment, either in itself or in its method of operation, shall not constitute a hazard to the safety of life or property.

§ 97.103 Station log requirements.

An accurate legible account of station operation shall be entered into a log for each amateur radio station. The following items shall be entered as a minimum:

(a) The call sign of the station, the signature of the station licensee, or a photocopy of the station license.

(b) The locations and dates upon which fixed operation of the station was initiated and terminated. If applicable, the location and dates upon which portable operation was initiated and terminated at each location.

(1) The date and time periods the duty control operator for the station was other than the station licensee, and the signature and primary station call sign of that duty control operator.

(2) A notation of third party traffic sent or received, including names of all third parties, and a brief description of the traffic content. This entry may be in a form other than written, but one which can be readily transcribed by the licensee into written form.

(3) Upon direction of the Commission, additional information as directed shall be recorded in the station log.

§ 97.105 Retention of logs.

The station log shall be preserved for a period of at least 1 year following the last date of entry and retained in the possession of the licensee. Copies of the log, including the sections required to be transcribed by § 97.103, shall be available to the Commission for inspection.

EMERGENCY OPERATIONS

§ 97.107 Operation in emergencies.

In the event of an emergency disrupting normally available communication facilities in any widespread area or areas, the Commission, in its discretion, may declare that a general state of communications emergency exists, designate the area or areas concerned, and specify the amateur frequency bands, or segments of such bands, for use only by amateurs participating in emergency communication within or with such affected area or areas. Amateurs desiring to request the declaration of such a state of emergency should communicate with the Commission's Engineer in Charge of the area concerned. Whenever such declaration has been made, operation of and with amateur stations in the area concerned shall be only in accordance with the requirements set forth in this section, but such requirements shall in nowise affect other normal amateur communication in the affected area when conducted on frequencies not designated for emergency operation.

(a) All transmissions within all designated amateur communications bands[1] other than communications relating directly to relief work, emergency service, or the establishment and maintenance of efficient amateur radio networks for the handling of such communications shall be suspended. Incidental calling, answering, testing or working (including casual conversations, remarks or messages) not pertinent to constructive handling of the emergency situation shall be prohibited within these bands.

[1] The frequency 4383.8 kHz may be used by any station authorized under this part to communicate with any other station in the State of Alaska for emergency communications. No airborne operations will be permitted on this frequency. Additionally, all stations operating on this frequency must be located in or within 50 nautical miles of the State of Alaska.

(b) The Commission may designate certain amateur stations to assist in the promulgation of information relating to the declaration of a general state of communications emergency, to monitor the designated amateur emergency communications bands, and to warn non-complying stations observed to be operating in those bands. Such station, when so designated, may transmit for that purpose on any frequency or frequencies authorized to be used by that station, provided such transmission do not interfere with essential emergency communications in progress; however, such transmissions shall preferably be made on authorized frequencies immediately adjacent to those segments of the amateur bands being cleared for the emergency. Individual transmissions for the purpose of advising other stations of the existence of the communications emergency shall refer to this section by number (§ 97.107) and shall specify, briefly and concisely, the date of the Commission's declaration, the area and nature of the emergency, and the amateur frequency bands or segments of such bands which constitute the amateur emergency communications bands at the time. The designated stations shall not enter into discussions with other stations beyond furnishing essential facts relative to the emergency, or acting as advisors to stations desiring to assist in the emergency, and the operators of such designated stations shall report fully to the Commission the identity of any stations failing to comply, after notice, with any of the pertinent provisions of this section.

(c) The special conditions imposed under the provisions of this section shall cease to apply only after the Commission, or its authorized representative, shall have declared such general state of communications emergency to be terminated: however, nothing in this paragraph shall be deemed to prevent the Commission from modifying the terms of its declaration from time to time as may be necessary during the period of a communications emergency, or from removing those conditions with respect to any amateur frequency band or segment of such band which no longer appears essential to the conduct of the emergency communications.

OPERATION OF ADDITIONAL STATIONS

§ 97.109 Operation of a control station.

(a) Amateur frequency bands above 220 MHz, excepting 435 to 438 MHz, may be used for emissions by a control station. Frequencies below 225 MHz used for control links must be monitored by the control operator immediately prior to, and during, periods of operation.

(b) Where a remotely controlled station has been authorized to be operated from one or more remote control stations, those remote control stations may be operated either mobile or portable.

§ 97.110 Operation of an auxiliary link station.

(a) An auxiliary link station may use amateur frequency bands above 220 MHz excepting 435 to 438 MHz for emissions. Frequencies below 225 MHz used by an auxiliary link station shall be monitored by the control operator immediately prior to, and during, periods of operation.

(b) An auxiliary link station may only be used for fixed operation from the location specified on the station license, and only when its associated station(s) is operated from its authorized land location.

§ 97.111 Operation of a repeater station.

(a) Emissions from a repeater station shall be discontinued within 5 seconds after cessation of radiocommunication by the user station. Provisions to automatically limit the access to a repeater station may be incorporated, but are not mandatory.

(b) The transmitting and receiving frequencies utilized by the repeater station shall be continuously monitored by the control operator immediately prior to, and during, periods of operation.

(c) A repeater station may be concurrently operated on more than one frequency band. Crossband operation of repeater stations is prohibited, i.e., both input (receiving) and output (transmitting) frequencies for a particular repeated transmission must be within the same frequency band. Operation on more than one output frequency on a single frequency band is prohibited except when specifically approved by the Commission. Repeater stations authorized to operate in conjunction with one or more auxiliary link stations may utilize an input frequency in a different frequency band provided the input frequency of the auxiliary link station(s) is in the same frequency band as the output frequency of the repeater station.

(d) A repeater station shall be operated in a manner so as to assure that the station is not used for one-way radiocommunication other than provided for in § 97.91.

(e) A station licensed as a repeater station may only be operated as a repeater station, excepting for short periods for testing or for emergencies.

(f) When in operation, the log of a repeater station must also show the following information for each frequency band in use.

(1) Location of the station transmitting antenna, marked upon a topographic map having a scale of 1:250,000, and contour intervals.[1]

(2) The transmitting antenna height above average terrain.[2]

(3) The effective radiated power in the horizontal plane for the main lobe of the antenna pattern, calculated for maximum transmitter output power.

(4) The transmitter output power.

(5) The loss in the transmission line between the transmitter and the antenna, expressed in decibels.

(6) The relative gain in the horizontal plane of the transmitting antenna.

(7) The horizontal and vertical radiation patterns of the transmitting antenna, with reference to true north (for horizontal pattern only), expressed as relative field strength (voltage) or in decibels, drawn upon polar coordinate graph paper, and method of determining the patterns.

SUBPART E—PROHIBITED PRACTICES AND ADMINISTRATIVE SANCTIONS

Prohibited Transmissions and Practices

§ 97.112 No remuneration for use of station.

(a) An amateur station shall not be used to transmit or receive messages for hire, nor for communication for material compensation, direct or indirect, paid or promised.

(b) Control operators of a Club Station may be compensated when the club station is operated primarily for the purpose of conducting amateur radiocommunication to provide telegraphy practice transmissions intended for persons learning or improving proficiency in the International Morse Code, or to disseminate information bulletins consisting solely of subject matter having direct interest to the Amateur Radio Service provided:

(1) The station conducts telegraphy practice and bulletin transmission for at least 40 hours per week;

(2) The station schedules operations on all allocated medium and high frequency amateur bands using reasonable measures to maximize coverage.

(3) The schedule of normal operating times and frequencies is published at least 30 days in advance of the actual transmissions.

Control operators may accept compensation only for such periods of time during which the station is transmitting telegraphy practice or bulletins. A control operator shall not accept any direct or indirect compensation for periods during which the station is transmitting material other than telegraphy practice or bulletins.

§ 97.113 Broadcasting prohibited.

Subject to the provisions of § 97.91, an amateur station shall not be used to engage in any form of broadcasting, that is, the dissemination of radio communications intended to be received by the public directly or by the intermediary of relay stations, nor for the retransmission by automatic means of programs or signals emanating from any class of station other than amateur. The foregoing provisions shall not be construed to prohibit amateur operators from giving their consent to the rebroadcast by broadcast stations of the transmissions of their amateur stations, provided, that the transmissions of the amateur stations shall not contain any direct or indirect reference to the rebroadcast.

§ 97.114 Third party traffic.

The transmission or delivery of the following amateur radiocommunication is prohibited:

(a) International third party traffic except with countries which have assented thereto;

(b) Third party traffic involving material compensation, either tangible or intangible, direct or indirect, to a third party, a station licensee, a control operator, or any other person.

(c) Except for an emergency communication as defined in this part, third party traffic consisting of business communications on behalf of any party. For the purpose of this section business communication shall mean any transmission or communication the purpose of which is to facilitate the regular business or commercial affairs of any party.

§ 97.115 Music prohibited.

The transmission of music by an amateur station is forbidden.

§ 97.116 Amateur radiocommunication for unlawful purposes prohibited.

The transmission of radiocommunication or messages by an amateur radio station for any purpose, or in connection with any activity, which is contrary to Federal, State, or local law is prohibited.

§ 97.117 Codes and ciphers prohibited.

The transmission by radio of messages in codes or ciphers in domestic and international communications to or between amateur stations is prohibited. All communications regardless of type of emission employed shall be in plain language except that generally recognized abbreviations established by regulation or custom and usage are permissible as are any other abbreviations or signals where the intent is not to obscure the meaning but only to facilitate communications.

---

[1] Indexes and ordering information for suitable maps are available from U.S. Geological Survey, Washington, D.C. 20242, or Federal Center, Denver, Colorado 80225.

[2] See Appendix 5.

§ 97.119 **Obscenity, indecency, profanity.**

No licensed radio operator or other person shall transmit communications containing obscene, indecent, or profane words, language, or meaning.

§ 97.121 **False signals.**

No licensed radio operator shall transmit false or deceptive signals or communications by radio, or any call letter or signal which has not been assigned by proper authority to the radio station he is operating.

§ 97.123 **Unidentified communications.**

No licensed radio operator shall transmit unidentified radio communications or signals.

§ 97.125 **Interference.**

No licensed radio operator shall willfully or maliciously interfere with or cause interference to any radio communication or signal.

§ 97.127 **Damage to apparatus.**

No licensed radio operator shall willfully damage, or cause or permit to be damaged, any radio apparatus or installation in any licensed radio station.

§ 97.129 **Fraudulent licenses.**

No licensed radio operator or other person shall obtain or attempt to obtain, or assist another to obtain or attempt to obtain, an operator license by fraudulent means.

ADMINISTRATIVE SANCTIONS

§ 97.131 **Restricted operation.**

(a) If the operation of an amateur station causes general interference to the reception of transmissions from stations operating in the domestic broadcast service when receivers of good engineering design including adequate selectivity characteristics are used to receive such transmissions and this fact is made known to the amateur station licensee, the amateur station shall not be operated during the hours from 8 p.m. to 10:30 p.m., local time, and on Sunday for the additional period from 10:30 a.m. until 1 p.m., local time, upon the frequency or frequencies used when the interference is created.

(b) In general, such steps as may be necessary to minimize interference to stations operating in other services may be required after investigation by the Commission.

§ 97.133 **Second notice of same violation.**

In every case where an amateur station licensee is cited within a period of 12 consecutive months for the second violation of the provisions of §§ 97.61, 97.63, 97.65, 97.71, or 97.73, the station licensee, if directed to do so by the Commission, shall not operate the station and shall not permit it to be operated from 6 p.m. to 10:30 p.m., local time, until written notice has been received authorizing the resumption of full-time operation. This notice will not be issued until the licensee has reported on the results of tests which he has conducted with at least two other amateur stations at hours other than 6 p.m. to 10:30 p.m., local time. Such tests are to be made for the specific purpose of aiding the licensee in determining whether the emissions of the station are in accordance with the Commission's rules. The licensee shall report to the Commission the observations made by the cooperating amateur licensees in relation to the reported violations. This report shall include a statement as to the corrective measures taken to insure compliance with the rules.

§ 97.135 **Third notice of same violation.**

In every case where an amateur station licensee is cited within a period of 12 consecutive months for the third violation of §§ 97.61, 97.63, 97.65, 97.71, or 97.73, the station licensee if directed by the Commission, shall not operate the station and shall not permit it to be operated from 8 a.m. to 12 midnight, local time, except for the purposes of transmitting a prearranged test to be observed by a monitoring station of the Commission to be designated in each particular case. The station shall not be permitted to resume operation during these hours until the licensee is authorized by the Commission, following the test, to resume full-time operation. The results of the test and the licensee's record shall be considered in determining the advisability of suspending the operator license or revoking the station license, or both.

§ 97.137 **Answers to notices of violations.**

Any licensee receiving official notice of a violation of the terms of the Communications Act of 1934, as amended, any legislative act, Executive order, treaty to which the United States is a party, or the rules and regulations of the Federal Communications Commission, shall, within 10 days from such receipt, send a written answer direct to the office of the Commission originating the official notice: *Provided, however*, That if an answer cannot be sent or an acknowledgment made within such 10-day period by reason of illness or other unavoidable circumstances, acknowledgment and answer shall be made at the earliest practicable date with a satisfactory explanation of the delay. The answer to each notice shall be complete in itself and shall not be abbreviated by reference to other communications or answers to other notices. If the notice relates to some violation that may be due to the physical or electrical characteristics of transmitting apparatus, the answer shall state fully what steps, if any, are taken to prevent future violations, and if any new apparatus is to be installed, the date such apparatus was ordered, the name of the manufacturer, and promised date of delivery. If the notice of violation relates to some lack of attention or improper operation of the transmitter, the name of the operator in charge shall be given.

SUBPART F—RADIO AMATEUR CIVIL EMERGENCY SERVICE (RACES)

GENERAL

§ 97.161 **Nature of this service.**

(a) The Radio Amateur Civil Emergency Service provides for amateur radio operation for civil defense communications purposes only, during periods of local, regional or national civil emergencies, including any emergency which may necessitate invoking of the President's War Emergency Powers under the provisions of Section 606 of the Communications Act of 1934, as amended.

(b) Pursuant to the provisions of section 4(j) of the Communications Act of 1934, as amended, records relating to the Radio Amateur Civil Emergency Service shall not be open to general public inspection.

§ 97.163 **Definitions.**

For the purposes of this subpart, the following definitions are applicable:

(a) *Radio Amateur Civil Emergency Service.* A radiocommunication service carried on by licensed amateur radio stations while operating on specifically designated segments of the regularly allocated amateur frequency bands under the direction of authorized local, regional or federal civil defense officials pursuant to an approved civil defense communications plan.

(b) *Radio Amateur Civil Emergency Station.* An amateur radio station which is authorized to operate in the Radio Amateur Civil Emergency Service for the purpose of transmitting and receiving civil defense communications.

(c) *Civil defense communications.* Communications or signals essential to the conduct of civil defense activities of duly authorized civil defense organizations, including communications directly concerning safety of life, preservation of property, maintenance of law and order, alleviation of human suffering and need and dissemination of warnings of enemy attack to the civilian population in case of actual or impending armed attack or in any disaster or other incident endangering the public welfare. Such communications may also include transmissions necessary to establishment and maintenance of the radio system and communications essential to the training of civil defense personnel.

(d) *Civil defense authority.* The legally appointed Director of Civil Defense, or his authorized alternate or representative, for the particular geographical area (city, county, etc.) which a proposed radio station is intended to serve, and who is responsible to local governmental authority for protection and aid to the civilian population in the event of armed attack or of any disaster or other incident endangering public safety.

(e) *Civil Defense Communications Officer.* The official of any duly constituted civil defense organization having direct responsibility under the Director of that organization for the provision, organization, maintenance, readiness, and utilization of all means of communication to be used by such civil defense organization in the performance of its lawful functions.

(f) *Civil Defense Radio Officer.* The duly designated official of a legally constituted civil defense organization who is directly responsible either to the Communications Officer or to the Director of such civil defense organization for the provision, organization, maintenance, readiness, and utilization of radio communications facilities for civil defense use.

(g) *Radio Amateur Civil Emergency Network.* All radio amateur civil emergency stations intended to be included in the civil defense communications plan of the area concerned and which operate, or are to operate, in conjunction with a single control station. Such network may be made up of several separately authorized radio amateur civil emergency stations or units of such stations, or may be made up of several units of the same station operated at different locations. In addition, the same radio amateur civil emergency station or any unit of such station may be a part of more than one network; e.g., the control station of one network may also be the control station or a member station of another network operated in conjunction therewith.

(h) *Net control station.* Any authorized Radio Amateur Civil Emergency Station unit designated by the civil defense radio officer, with the approval of the Director of Civil Defense or the Civil Defense Communications Officer, to direct the use and operation of other station units of the same Radio Amateur Civil Emergency Network.

(i) *Civil defense communications plan.* The plan under which communications facilities are provided to all branches and phases of the civil defense organization in the area concerned and for all of its activities. Such plan may be drawn up in accordance with the needs of the particular area affected and the facilities, including licensed radio operators and stations, available in that particular area. Plans need not be uniform, but to be acceptable to the Commission they must comply with the following:

(1) The plan must be clearly described in writing, and it may include diagrams and sketches. It must include a general description of the facilities and personnel available to provide communications for civil defense purposes and the expected usage to be made thereof.

(2) The plan must have been approved by the state and federal civil defense authorities having jurisdiction of the area affected.

(3) The plan must include the name, address, official title, and a statement of the qualifications of the Civil Defense Radio Officer (and of any and all alternate Radio Officers) responsible for the organization, training, and utilization of the radio amateur civil emergency station networks under that plan, and the name, address, and official title of the civil defense official responsible for the coordination of all civil defense activities of the area concerned.

(4) The plan must include a general description of each radio amateur civil emergency station network under the jurisdiction of each respective Civil Defense Radio Officer, showing location of fixed installations, purpose, area of activity to be served, an estimate of the number of radio amateur stations and independent operating units of such stations intended to be used in the network, and a description, including the location and call sign, of its control station and any alternate control station or stations.

(5) The plan must include a general statement as to the frequency bands to be used by the radio amateur civil emergency station networks and the approximate number of stations, or units of such stations, to be operated in each such band, together with a description of the method which has been adopted for liaison and coordination of frequency usage with other similar networks in the same and adjacent areas.

(6) The plan must include a statement setting forth the facilities available to the area and the procedures to be followed in determining the loyalty and general reliability of all civil defense Radio Officers, amateur radio station licensees and radio operators intended to be utilized in the implementation of that plan. (See §§ 97.173(b), 97.175(c), and 97.203(a).)

§ 97.165 Applicability of rules governing amateur radio stations and operators.

In all cases not specifically covered by the regulations contained in this subpart, licensed amateur stations authorized to be operated in the Radio Amateur Civil Emergency Service shall be governed by the provisions of the rules governing amateur radio stations and operators (Subparts A through E of this part) which are not in conflict herewith. In any case of conflict, the rules governing the Radio Amateur Civil Emergency Service shall govern in respect to any station operated in that service.

ORGANIZATION

§ 97.167 Organization of networks.

To supplement or extend other means of communication available to the civil defense organization or to provide necessary communications for which no other means exist, local radio amateur civil emergency station networks shall be organized by the civil defense authority of the area concerned and under the immediate direction of the Civil Defense Radio Officer. Such networks shall include all licensed amateur radio stations which are intended to be included in the civil defense communications plan of the area concerned. In any particular area there may be several such networks and each network may be independent of the others. Whenever there is more than one network in the same area, all such networks must share, under a single civil defense communications plan, the available frequencies in an efficient and orderly manner. The various networks in adjacent areas shall establish proper liaison and a description of the arrangements made shall become a part of their respective civil defense communications plans. Such arrangements shall provide for the efficient sharing of frequencies, plans for operating procedure designed to avoid mutual interference, and the exchange of communications facilities upon an inter-area basis where need for such exchange may arise.

§ 97.169 Approval of civil defense communications plans.

(a) All civil defense communications plans which provide for the utilization of radio amateur civil emergency stations for civil defense purposes must be submitted to and approved by the responsible state (or territorial) and federal civil defense authorities before the licensed amateur stations intended to be used will be authorized to operate in the radio amateur civil emergency service.

(b) Material changes or modifications in such civil defense communications plans which alter the basic information required shall be submitted for approval in the same manner as the original plans.

(c) Written certification of approval by the competent state and federal civil defense authorities of each civil defense communications plan, or of any changes or modifications thereof, shall accompany the copies of such plans, changes, or modifications which are submitted to the Commission in accordance with the provisions of this part.

§ 97.171 Certification of Civil Defense Radio Officer.

(a) Certification of the Civil Defense Radio Officer shall be made on FCC Form 482. Such form shall be executed by the civil defense authority responsible for the coordination of all civil defense activities of the area concerned and show:

(1) The name, address, and area of responsibility of such civil defense radio officer,

(2) statement by him that he has accepted such appointment and agrees to perform faithfully the duties of that office, including those prescribed by this subpart,

(3) a certification by the responsible civil defense authority that he has satisfied himself that the named civil defense radio officer is fully qualified in accordance with the provisions of § 97.173, and

(4) the effective date of the appointment of the civil defense radio officer and the name of any previous civil defense radio officer whose appointment is terminated.

(b) FCC Form 482, when completed in accordance with this section, shall be forwarded to the Commission via the responsible state and federal civil defense officials whose approval (or disapproval) shall be clearly indicated on the form.

§ 97.173 Qualifications of Civil Defense Radio Officer.

No person shall be considered qualified as a Civil Defense Radio Officer until he shall have been found to satisfy the following minimum requirements:

(a) He shall hold either (1) a valid commercial radio operator's license of either first or second class (radiotelegraph or radiotelephone) issued by the Commission, or (2) a valid amateur operator license issued by the Commission, other than the Technician or Novice Class.

(b) A determination shall have been made as to his loyalty to the United States and his general reliability, in accordance with the procedures provided in the approved civil defense communications plan of the area concerned.

(c) It shall have been determined that his technical and administrative qualifications are adequate for the proper performance of his duties.

§ 97.175 Duties of Civil Defense Radio Officer.

The duties of the Civil Defense Radio Officer shall include among such other duties as may be assigned or as may be required in accordance with the provisions of this subpart.

(a) The direction and supervision of all radio stations forming the radio amateur civil emergency networks in accordance with the approved civil defense communications plan for the area involved.

(b) Provision for adequate monitoring of all transmissions of the stations under his supervision to assure compliance with the rules and regulations of the Commission, and to guard against improper use of the radio stations and intentional or inadvertent transmissions which might jeopardize the defense or security of the United States.

(c) The recommendation to the Commission for the granting of authorizations to individual amateurs for operation in this service, and certification to the Commission as to the loyalty to the United States and reliability of such individuals and the certification required in accordance with § 97.181.

(d) The recommendation to the Commission for cancellation of any authorization previously recommended or certified whenever subsequent investigation or circumstances indicate that the original recommendation or certification should not have been made.

STATION AUTHORIZATIONS

§ 97.177 Station authorization required.

No radio station may be operated in the Radio Amateur Civil Emergency Service except pursuant to an authorization for such operation issued by the Federal Communications Commission.

§ 97.179 Eligibility for station authorization.

An authorization to operate a station in the Radio Amateur Civil Emergency Service will be issued only to a person who holds an amateur radio operator license, other than Technician or Novice Class, and an appropriate amateur radio station license.

§ 97.181 Filing of application.

Each application for a station authorization or for renewal thereof shall be submitted on FCC Form 481-1, signed by the applicant and countersigned by the appropriate Civil Defense Radio Officer, who shall certify to the following:

(a) That the applicant has satisfied all requirements (both local and federal) for participation in the civil defense organization and is actually enrolled as a member of the local organization which serves the area where the station will operate.

(b) That the amateur station licensed in the name of the applicant has been approved for and, when authorized by the Commission, will actually constitute a unit of a civil defense communications network in accordance with an approved civil defense communications plan or amendment thereof.

§ 97.183 Additional data required.

Each application for a station authorization in the Radio Amateur Civil Emergency Service shall be accompanied by the following data unless such material has already been submitted to the Commission, in which case the application shall clearly identify the material previously submitted:

(a) A copy of the approved communications plan (as defined in this part) for the civil defense communications network in which the station will operate, together with a copy of each approved amendment, change or modification of that plan.

(b) The official certification of the Civil Defense Radio Officer as provided in this subpart.

§ 97.185 Single application for all equipment under one amateur station license.

Only one application need be filed for any one amateur station, including all transmitting equipment under the control of the licensee of that station, even though individual units of such station are capable of being operated and are intended to be operated independently at different locations, or as portable or mobile stations with no fixed locations. No distinction need be made between those units which are personally owned by the amateur station licensee and those units which are otherwise under his technical control for operation in this service.

§ 97.187 Issuance of station authorization.

An authorization to operate in this service will be issued in the discretion of the Commission upon satisfactory completion of all requirements of this subpart and proper certification that the requirements of the civil defense organization for which the station will be used have been or are being complied with. The station authorization (Form 481-3) will be forwarded to the Civil Defense Radio Officer for delivery to the applicant. Such authorization will be accompanied by a stub (Form 481-2) which may be retained by the civil defense radio officer for his records.

§ 97.189 Term of station authorization.

(a) Authorization to operate an amateur station in the Radio Amateur Civil Emergency Service will be issued for a term running concurrently with the term of the amateur radio station license. Application for renewal of such authorization shall be filed concurrently with application for renewal of the basic amateur radio station license.

(b) Whenever, under rules contained in Subparts A through E of this part, modification of the basic amateur station license becomes necessary, if such modification affects the information submitted with the original application for authorization in the Radio Amateur Civil Emergency Service, application for modification of the Radio Amateur Civil Emergency service station authorization shall be submitted concurrently therewith.

(c) Nothing in this section shall be construed to alter the Commission's authority to cancel or amend a station authorization in the Radio Amateur Civil Emergency Service in accordance with the applicant's agreement as indicated on the initial application for station authorization.

§ 97.191 Cancellation of station authorization.

(a) Each authorization for operation in the Radio Amateur Civil Emergency Service shall be issued with the express provision that such authorization is subject to revocation or cancellation without hearing whenever, in the opinion of the Commission, the security of the United States or the proper functioning of the Radio Amateur Civil Emergency Service would be served thereby.

(b) The station authorization shall be submitted to the Commission (via the Civil Defense Radio Officer) for cancellation under the following circumstances:

(1) The station for which the authorization was issued becomes inactive for a period of three months or it is not planned to use the station in the radio amateur civil emergency network for a period of at least three months.

(2) The basic amateur radio station license of the station has expired and has not been renewed.

(3) In cases where the amateur radio station license and the radio amateur civil emergency station authorization have both been modified, the original authorization of the latter shall be submitted to the Commission immediately upon receipt by the licensee of a new or modified authorization.

TECHNICAL REQUIREMENTS

§ 97.193 Frequencies available.

(a) Except as provided in paragraph (e) of this section, the following frequency and frequency bands and associated emissions are available on a nonexclusive basis to the individual class of stations or units of such stations in the Radio Amateur Civil Emergency Service.

(1) For use only by authorized stations or units of such stations which are operated under the direct supervision of duly designated and responsible officials of the civil defense organization:

Frequency band:	*Authorized emission*
1800–1825 kHz [1]	0.1A1, 1.1F1, 6A3
1975–2000 kHz [1]	0.1A1, 1.1F1, 6A3
3500–3510 kHz	0.1A1, 1.1F1
3990–4000 kHz	0.1A1, 1.1F1, 6A3, 6F3

[1] Use of frequencies in the band 1800–2000 kHz is subject to the priority of the Loran system of radionavigation in this band and to the geographical, frequency, emission, and power limitations contained in § 97.61 of the rules governing amateur radio stations and operators (Subparts A through E of this part). The use of these frequencies by stations authorized to be operated in the Radio Amateur Civil Emergency Service shall not be a bar to expansion of the radionavigation (Loran) service, and such use shall be considered temporary in the sense that it shall remain subject to cancellation or to revision, in whole or in part, without hearing, whenever the Commission shall deem such cancellation or revision to be necessary or desirable in the light of the priority within this band of the Loran system of radionavigation.

(2) For use by all authorized stations only in the continental United States, except that, the bands 7245–7255 and 14.220–14.230 kHz are also available in Alaska, Hawaii, Puerto Rico, and the Virgin Islands:

Frequency band:	*Authorized emission*
3510–3516 kHz	0.1A1 1.1F1.
3516–3550 kHz [1]	0.1A1, 1.1F1.
3984–3990 kHz	0.1A1, 1.1F1, 6A3, 6F3.
7097–7103 kHz	0.1A1, 1.1F1.
7103–7125 kHz [1]	0.1A1, 1.1F1.
7245–7255 kHz [1]	0.1A1, 1.1F1, 6A3, 6F3.
14047–14053 kHz	0.1A1, 1.1F1.
14220–14230 kHz [1]	0.1A1, 1.1F1, 6A3, 6F3.
21047–21053 kHz	0.1A1, 1.1F1.

[1] The availability of the frequency band 3516–3550 kHz, 7103–7125 kHz, 7245–7247 kHz, 7253–7255 kHz, 14220–14222 kHz and 14228–14230 kHz for use during periods of actual civil defense emergency is limited to the initial 30 days of such emergency, unless otherwise ordered by the Commission.

(3) For use by all authorized stations:

Frequency or frequency bands:	*Authorized emission*
3997 kHz [1]	0.1A1, 6A3.
28.55–28.75 MHz	0.1A1, 6A3, 6F3, 6A4.
29.45–29.65 MHz	0.1A1, 1.1F1, 6A3, 6A4, 40F3.

50.35–50.75 MHz	0.1A1, 6A2, 6F2, 6A3, 6F3, 6A4.
53.30 MHz [1]	40F3.
53.35–53.75 MHz	0.1A1, 1.1F1, 6A2, 6F2, 6A3, 6A4, 40F3.
145.17–145.71 MHz.	0.1A1, 1.1F1, 6A2, 6F2, 6A3, 6A4, 40F3.
146.79–147.33 MHz.	0.1A1, 1.1F1, 6A2, 6F2, 6A3, 6A4, 40F3.
220–225 MHz	0.1A1, 1.1F1, 6A2, 6F2, 6A3, 6A4, 40F3.

[1] For use in emergency areas when required to make initial contact with military units; also, for communication with military stations on matters requiring coordination.

(b) The selection and use of specific frequencies within the authorized frequency bands by stations in the Radio Amateur Civil Emergency Service shall be in accordance with a coordinated local area and adjacent area civil defense communications plan and applicable rules of this part.

(c) Except as provided in paragraph (d) of this section, at such time as any or all of these frequency bands are withdrawn from availability to stations operating in the Amateur Radio Service, such bands shall be jointly available to stations in the Radio Amateur Civil Emergency Service and to stations in the military services for training and tactical operations. At that time, in areas where interference might occur, local mutual arrangements shall be made regarding times of operation such as to preclude or satisfactorily alleviate interference. In time of actual civil defense emergency, stations in the Radio Amateur Civil Emergency Service shall have absolute priority.

(d) In the band 220 to 225 MHz, stations operating in the Radio Amateur Civil Emergency Service shall not at any time cause harmful interference to the government radiolocation service.

(e) A repeater station in the Radio Amateur Civil Emergency Service may operate on any frequency, and with any associated emission, above 50 MHz listed in paragraph (a) of this section, except for 220 MHz to 222 MHz.

§ 97.195 Classification of emissions.

(a) For the purposes of this subpart, the authorized emissions, as contained in the table of § 97.193, are defined as follows:

0.1A1—Continuous wave telegraphy.
1.1F1—Frequency shift telegraphy.
6A2—Telegraphy amplitude modulated at audio frequency.
6F2—Telegraphy frequency modulated at audio frequency.
6A3—Commercial quality amplitude modulated telephony.
6F3—Narrow band frequency or phase modulated telephony.
40F3—Wide band frequency or phase modulated telephony.
6A4—Amplitude modulated facsimile.

(b) On frequencies where wide band frequency or phase modulated telephony (40F3) is authorized, narrow band frequency or phase modulated telephony (6F3) may also be employed; similarly, where commercial quality amplitude modulated telephony (6A3) is authorized, single or double sideband amplitude modulated telephony, with or without carrier or with reduced carrier, may also be employed.

§ 97.197 Transmitter power.

The transmitting equipment of a radio station in this service shall be adjusted in such manner as to produce the minimum radiation necessary to carry out the communications desired. No station operating in this service shall use a direct current plate power input to the vacuum tube or tubes supplying energy to the antenna in excess of that permitted to be used by a licensed amateur radio station when operated on the same frequencies or in the same frequency bands in accordance with the provisions of the rules governing amateur radio stations and operators (Subparts A through E of this part).

§ 97.199 Equipment requirements.

(a) Except under the conditions specified in paragraph (b) of this section, all stations authorized to be operated in the Radio Amateur Civil Emergency Service shall be capable of receiving on the same frequencies or frequency bands utilized for transmission.

(b) When a station in this service is operated only on a single frequency or frequency band for cross-band operation in communication with a station or stations operating on another frequency or in another frequency band, or in other services, such station shall be capable of receiving the station with which it is communicating.

(c) The direct modulation of an oscillator with a frequency stability less than that obtainable with crystal control, or the radiation of a signal having simultaneous amplitude and frequency or phase modulation, is prohibited on frequencies below 220 MHz.

§ 97.201 Alleviation of harmful interference.

(a) When emissions of stations in the Radio Amateur Civil Emergency Service, other than those necessary to carry on the desired communications, cause harmful interference to stations in this or any other service, the Commission may, in its discretion, require appropriate technical changes in the equipment to alleviate the interference.

(b) When the emissions of stations in the Radio Amateur Civil Emergency Service that are necessary to carry on the desired communications cause harmful interference to stations in other radio services, appropriate action shall be taken to alleviate such interference including, if necessary, the suspension (except during times of an actual state of civil emergency) of such emissions as cause the interference.

Operating Requirements

§ 97.203 Operator requirements.

(a) No person shall operate a station in the Radio Amateur Civil Emergency Service unless (1) that person holds a valid radio operator license of the proper grade, as described in this section, and (2) that person holds a valid written certification by the chief of the local, regional, or state Civil Defense organization of the area in which he serves that he has satisfied all federal, state, and local requirements for enrollment in the Civil Defense organization as a radio operator and is actually enrolled therein. Such certification shall clearly indicate that a determination has been made as to his loyalty to the United States and general reliability in accordance with the procedures described in the approved civil defense communications plan for the area concerned. (See §§ 97.163(i) and 97.169.)

(b) The person manipulating the key of a manually operated radiotelegraph transmitter of a station authorized to operate in this service shall hold either (1) any class of amateur operator license issued by the Commission, other than the Technician or Novice Class, or (2) any class of commercial radiotelegraph operator license issued by the Commission other than the Temporary Limited Radiotelegraph Second Class Operator License, together with the certification required in accordance with the provisions of paragraph (a) of this section.

(c) Except as specifically provided in paragraphs (a) and (b) of this section, any station in the Radio Amateur Civil Emergency Service may be operated by the holder of any class of amateur or commercial radio operator license issued by the Commission other than a Temporary Limited Radiotelegraph Second Class Operator License or an Aircraft Radiotelephone Operator Authorization: *Provided*, That, when such operation is performed by the holder of a Novice Class amateur operator license or by the holder of a commercial radiotelephone or radiotelegraph third class operator license or restricted operator permit; (1) such operator shall be prohibited from making any adjustments that may result in improper transmitter operation, (2) the equipment shall be so designed and installed that none of the operations necessary to be performed during the course of the normal rendition of the service of the station may cause off-frequency operation or result in any unauthorized radiation, and (3) any needed adjustments of the transmitter that may affect the proper operation of the station shall be regularly made by or under the immediate supervision and responsibility of the holder of either an amateur operator license other than the Novice Class or a commercial radiotelephone or radiotelegraph first or second class operator license.

(d) All adjustments or tests during or coincident with the installation, servicing or maintenance of the transmitting equipment of a station in this service shall be made only by or under the immediate supervision and responsibility of the holder of either (1) an amateur operator license other than the Novice Class or (2) a commercial radiotelephone or radiotelegraph first or second class operator license issued by the Commission, who in addition holds the certification required in accordance with the provisions of paragraph (a) of this section.

§ 97.205 Operation at other than licensed location.

A station in this service, or any unit thereof, may be operated at any location in accordance with the approved civil defense communications plan for the area concerned, in the discretion of and as directed by the Civil Defense Radio Officer, without notice to the Commission and without limitation as to the length of time within which such operation takes place: *Provided*, That nothing in this section shall be construed to waive the necessity for modification of the authorization of a station in this service when the address of the licensee or the basic location of the station is changed, or for any other reason where, because of a change of the communications plan or other reason, the information heretofore furnished the Commission with the original application may be materially altered or changed.

§ 97.207 Availability of station authorizations and operator licenses.

(a) The original station authorization permitting operation of the licensed amateur station in the Radio Amateur Civil Emergency Service, or a photocopy thereof, shall be permanently attached to each transmitter of such station, including each transmitter which is capable of being operated and intended to be operated independently at different locations, if the transmitter is readily accessible, or, if the control position is located at a place other than the transmitter location, it may be posted at the control position: *Provided*, That, whenever a photocopy of the station authorization is utilized in compliance with the requirement of this paragraph, the original station authorization shall be made available for inspection upon reasonable request from any authorized representative of the Federal Government.

(b) The original radio operator license, or a verification card (FCC Form 758–F) in the case of the holder of a commercial radio operator license of the diploma type, of the operator controlling the emissions of a station authorized to be operated in this service together with the certification required by § 97.203(a), shall be carried on his person or kept immediately available at the place where he is operating the station or any independent unit of a station: *Provided*, That, whenever a verification card (FCC Form 758–F) is utilized in compliance with the requirement of this paragraph, the original operator license shall be made available for inspection upon reasonable request from an authorized representative of the Federal Government.

(c) When a licensed amateur station, or an independent unit of such station, is operated at a location other than that shown in its license in compliance with the provisions of this subpart, the basic amateur station license required by Subparts A through E of this part need not be readily available at the station or unit location, but shall be made available for inspection upon reasonable request from any authorized representative of the Federal Government.

§ 97.209 Radio station log.

(a) Except as otherwise expressly provided in this subpart, there shall be maintained at each radio amateur civil emergency station, or unit of such station, an accurate log of all operations. The following information shall be recorded in such station log:

(1) The name and address of the station licensee, the regularly assigned call sign of the station and unit number if any, the name of the radio amateur civil emergency network or networks in which the station is normally operated, and the d.c. plate power input to the vacuum tube or tubes supplying energy to the transmitting antenna system. This information need be entered only once in the log unless there is a change in any of the items specified in this subparagraph, but the original entry and each change shall show the date on which the entry was made.

(2) The date and time of beginning and end of each period during which the station was operated, the purpose of such operation, and the frequencies or bands of frequencies on which the operation took place.

(3) The call signs or other identification of all stations or units of such stations with which communications are established or attempted during such period of operation.

(4) The signature of the licensed operator on duty and in charge of the operation of the station or unit of such station during each period of operation, and the signature of each licensed operator who manipulated the key of any manually operated radiotelegraph transmitter of such station or unit. The signature of the operator shall be entered with the date and time at the beginning and end of each period during which he performed the foregoing duties, and at least once on each page additional to the first page, covering the period for which he was the responsible operator. The signatures of any additional operators who operate the transmitters(s) during the regular watch of another operator and details to indicate the periods during which they operated the transmitter(s) shall be entered in the proper form.

(5) Upon completion of each period of operation for any purpose, there shall be entered in the log a summary of such operation describing the nature thereof and, if message traffic or other record communications were exchanged with other stations, an estimate of the amount of such traffic handled together with a report on any unusual delays which were experienced in the delivery of such messages.

(6) There shall be no erasure, obliteration, or destruction of any part of the log of any station or station unit. Corrections shall be made by striking out the erroneous portion and initialing and dating the corrections.

(b) Mobile radio amateur civil emergency stations or station units, and portable radio amateur civil emergency stations or station units, where not being operated at pre-determined fixed locations, shall be exempt from the requirements of maintaining a log to the extent that the entries required under the preceding paragraph of this section are substantially contained in the log of another station or stations operating in the same radio amateur civil emergency networks. All stations or station units operating in accordance with the provisions of this subpart shall be exempt from the requirements concerning station logs contained in Subpart D of this part whenever it is shown that compliance with these requirements would interfere with the expeditious handling of civil defense communications or communication drills.

(c) The current portion of the log shall be kept at the location of the operating or control position of the station or unit. Other portions of the log shall be retained by the licensee for a period of one year, at a place determined by the civil defense Radio Officer to be appropriate and advisable: *Provided*, That the logs of a station in this service shall be made available for inspection upon reasonable request by any authorized representative of the Federal Government: *And provided further*, That those portions of any log covering operation of a station in this service in connection with any actual condition jeopardizing the public safety or affecting the national defense or security shall not be destroyed unless prior approval for such destruction shall have been received from the Commission.

§ 97.211 Station identification.

(a) Stations operating in the Radio Amateur Civil Emergency Service shall identify themselves in the same manner and under the same conditions as prescribed in Subpart D of this Part, except that:

(1) Additional designators to indicate portable or mobile operation, or to indicate operation at a location other than that specified in the station license, shall not be used.

(2) When engaged in network operation, after a station or unit has been fully identified at least once, further identification by that station or unit may be accomplished by the use of abbreviated call signs or other distinctive signals prescribed by the civil defense Radio Officer in lieu of the call signs otherwise required to be transmitted by that station or unit. A record of such abbreviated call signs or other distinctive signals shall be maintained by the Radio Officer and shall be made available for inspection upon reasonable request by any authorized representative of the Federal Government.

(b) When two or more separate units of a station, which is authorized to be operated in the Radio Amateur Civil Emergency Service, are operated independently at different locations, each unit shall separately identify itself by the addition of a unit number at the end of its call sign. When transmitting by telegraphy such additional identification shall immediately follow the basic call sign and to avoid confusion with portable or mobile indicators, shall not be separated therefrom by the use of the "slant" or fraction bar, or other punctuation mark or symbol.

§ 97.213 Tactical call signs.

Stations operating in this service, and independent units of such stations, may be assigned tactical or secret call signs by the Commission or by competent civil defense authority, and may utilize such tactical call signs in lieu of the call signs appearing on the station licenses when such use is directed by competent civil defense authority: *Provided*, That a list of all such tactical call signs assigned stations under his direction shall be maintained by the civil defense Radio Officer and shall be made available for inspection upon reasonable request by any authorized representative of the Federal Government: *And provided further*, That when such tactical call signs are intended to be used at times other than during communications in connection with actual or impending conditions which appear to jeopardize the defense or security of the United States, a list of such tactical call signs and the stations or units to which assigned shall be furnished the Commission prior to such use.

USE OF STATIONS

§ 97.215 Limitations on use of stations.

(a) No station authorized to be operated in this service other than a control station as defined in this subpart, shall be operated for the purpose of transmitting any signal, message, or other communications except with the permission and under the operational control of the control station of the network in which it is operating: *Provided*, That nothing in the foregoing shall be construed to prohibit the transmission by any station or unit of a station of such signals as may be necessary for the purpose of alerting or making contact with the control station of the network, or for the purpose of transmitting actual emergency civil defense communications if the control station is disabled or is otherwise inoperative.

(b) Nothing in this section shall be construed to prevent the operation of a station which is authorized to be operated in this service for the purpose of brief tests or adjustments during or coincident with the installation, servicing or maintenance of such station: *Provided*, That the transmissions of that station during such tests or adjustments shall not cause harmful interference to the conduct of communications by any other station.

(c) No station in this service shall be used to transmit or to receive messages for hire, nor to transmit communications for material compensation, direct or indirect, paid or promised.

§ 97.217 Hours of operation.

Stations in this service may be operated at such times and under such conditions as may be prescribed by the Communications Officer or other responsible official of the civil defense organization having jurisdiction over the area which the station will serve: *Provided*, That the communications of such stations shall at all times be in accordance with the permissible communications authorized in this subpart.

§ 97.219 Points of communication.

Stations in this service may communicate with each other, with stations in the Disaster Communications Service, and with stations of the United States Government which are authorized to exchange communications with stations in this service by the particular agency having control. In addition, stations in this service may communicate, for the purpose of exchanging civil defense communications, with any other station in any service provided by the Commission's rules, whenever such station is authorized to communicate with stations in the Radio Amateur Civil Emergency Service by the provisions of the Commission's rules governing the class of station concerned or in accordance with the provisions of § 2.405 of this chapter.

§ 97.221 Permissible communications.

Stations in this service are authorized to transmit only the following types of civil defense communications:

(a) Communications for training purposes consisting of necessary drills and tests to insure establishment and maintenance of orderly and efficient operation of the radio amateur civil emergency networks and such other radio stations and networks as may be associated therewith for the conduct of civil defense communications, including communications directly concerned with the conduct of practice alerts, practice blackouts, practice mobilization, and other comparable situations as may be ordered or initiated by competent civil defense authority or by the United States governmental or military authority charged with the defense of the area concerned. All messages which are transmitted in connection with such drills, exercises and tests shall be clearly identified as such by use of any one of the words "Drill" or "Exercise" or "Test" in the body of such messages.

(b) Communications when there is an impending or actual condition jeopardizing the public safety or affecting the national defense or security:

(1) Communications directly concerning the activation of the radio amateur civil emergency station networks or such other radio stations and networks as may be associated with the networks for the conduct of civil defense communications.

(2) Communications directly concerning the conduct of service by the radio amateur civil emergency networks and such other radio stations and networks as may be associated therewith.

(3) Communications directly concerning safety of life, preservation of property, maintenance of law and order, alleviation of human suffering and need, and combating of armed attack or sabotage.

(4) Communications directly concerning the accumulation and dissemination of public information or instructions to the civilian population essential to the activities of the civil defense organization or that of other authorized governmental or relief agencies.

(5) Communications directly concerning the transaction of business essential to public welfare.

§ 97.223 Use of codes and ciphers.

Any station in this service is authorized to transmit messages in codes and ciphers and to utilize any method of secret or coded authentication of its transmissions when such method of concealing the contents of messages or such authentication procedure is prescribed by the competent civil defense authority of the area served by the station and is approved by the cognizant federal civil defense authorities.

§ 97.225 Priority of communications.

The order of priority of communications by stations in this service, when there is an impending or actual condition jeopardizing the public safety or affecting the defense or security of an area, shall be determined by the cognizant civil defense authority of the area concerned or his authorized representative.

§ 97.227 Operating procedure.

The operating procedure, and the method of circuit control by the control station of each network, shall be determined by the responsible civil defense authority of the area concerned and shall, in general, conform as nearly as possible to the operating procedure normally followed in other services in the expeditious handling of message traffic by the method of transmission in use.

Subpart G—Operation of Amateur Radio Stations in the United States by Aliens

§ 97.301 Basis, purpose, and scope.

(a) The rules in this subpart are based on, and are applicable solely to, alien amateur operations pursuant to section 303(l)(2) and 310(a) of the Communications Act of 1934, as amended. (See Public Law 88–313, 78 Stat. 202.)

(b) The purpose of this subpart is to implement Public Law 88–313 by prescribing the rules under which an alien, who holds an amateur operator and station license issued by his government (hereafter referred to as an alien amateur), may operate an amateur radio station in the United States, in its possessions, and in the Commonwealth of Puerto Rico (hereafter referred to only as the United States).

§ 97.303 Permit required.

(a) Before he may operate an amateur radio station in the United States, under the provisions of sections 303(l)(2) and 310(a) of the Communications Act of 1934, as amended, an alien amateur licensee must obtain a permit for such operation from the Federal Communications Commission. A permit for such operation shall be issued only to an alien holding a valid amateur operator and station authorization from his government, and only when there is in effect a bilateral agreement between the United States and that government for such operation on a reciprocal basis by United States amateur radio operators.

§ 97.305 Application for permit.

(a) Application for a permit shall be made on FCC Form 610–A. Form 610–A may be obtained from the Commission's Washington, D.C., office, from any of the Commission's field offices and, in some instances, from United States missions abroad.

(b) The application form shall be completed in full in English and signed by the applicant. A photocopy of the applicant's amateur operator and station license issued by his government shall be filed with the application. The Commission may require the applicant to furnish additional information. The application must be filed by mail or in person with the Federal Communications Commission, Washington, D.C., 20554, U.S.A. To allow sufficient time for processing, the application should be filed at least 60 days before the date on which the applicant desires to commence operation.

§ 97.307 Issuance of permit.

(a) The Commission may issue a permit to an alien amateur under such terms and conditions as it deems appropriate. If a change in the terms of a permit is desired, an application for modification of the permit is required. If operation beyond the expiration date of a permit is desired, an application for renewal of the permit is required. In any case in which the permittee has, in accordance with the provisions of this subpart, made a timely and sufficient application for renewal of an unexpired permit, such permit shall not expire until the application has been finally determined. Application for modification or for renewal of a permit shall be filed on FCC Form 610–A.

(b) The Commission, in its discretion, may deny any

application for a permit under this subpart. If an application is denied, the applicant will be notified by letter. The applicant may, within 90 days of the mailing of such letter, request the Commission to reconsider its action.

(c) Normally, a permit will be issued to expire 1 year after issuance but in no event after the expiration of the license issued to the alien amateur by his government.

§ 97.309 Modification, suspension, or cancellation of permit.

At any time the Commission may, in its discretion, modify, suspend, or cancel any permit issued under this subpart. In this event, the permittee will be notified of the Commission's action by letter mailed to his mailing address in the United States and the permittee shall comply immediately. A permittee may, within 90 days of the mailing of such letter, request the Commission to reconsider its action. The filing of a request for reconsideration shall not stay the effectiveness of that action, but the Commission may stay its action on its own motion.

§ 97.311 Operating conditions.

(a) The alien amateur may not under any circumstances begin operation until he has received a permit issued by the Commission.

(b) Operation of an amateur station by an alien amateur under a permit issued by the Commission must comply with all of the following:

(1) The terms of the bilateral agreement between the alien amateur's government and the government of the United States;

(2) The provisions of this subpart and of Subparts A through E of this part;

(3) The operating terms and conditions of the license issued to the alien amateur by his government; and

(4) Any further conditions specified on the permit issued by the Commission.

(c) An alien amateur may operate on dates, at locations, or via an itinerary, significantly different from that specified in the application for his permit only under the condition that he has given advance notice of the particulars of such operation to the Commission in accordance with the requirements of § 97.95(a).

§ 97.313 Station identification.

(a) The alien amateur shall identity his station as follows:

(1) Radio telegraph operation: The amateur shall transmit the call sign issued to him by the licensing country followed by a slant (/) sign and the United States amateur call sign prefix letter(s) and number appropriate to the location of his station.

(2) Radiotelephone operation: The amateur shall transmit the call sign issued to him by the licensing country followed by the words "fixed", "portable" or "mobile", as appropriate, and the United States amateur call sign prefix letter(s) and number appropriate to the location of his station. The identification shall be made in the English language.

(b) At least once during each contact with another amateur station, the alien amateur shall indicate, in English, the geographical location of his station as nearly as possible by city and State, commonwealth, or possession.

Subpart H—Operation of Amateur Radio Stations in the United States by Permanent Resident Aliens

§ 97.401 Basis, purpose and scope.

(a) The rules in this subpart are based on and are applicable solely to those provisions of section 303(1)(3) and 310(a) of the Communications Act of 1934, as amended (see Public Law 92–81, 85 Stat. and 78 Stat. 202) whereby certain aliens admitted to the United States for permanent residence should be eligible to operate amateur radio stations and to hold licenses for their stations.

(b) The purpose of this subpart is to implement Public Law 92–81 by prescribing the rules under which an alien, who is a permanent resident of the United States and has filed a declaration of intention with a State or Federal court may operate an amateur radio station in the United States.

§ 97.403 License required.

(a) Before an alien, under Public Law 92–81, may operate an amateur radio station in the United States under the provisions of sections 303(1)(3) and 301(a) of the Communications Act of 1934, as amended, he must obtain a license for such operation from the Federal Communications Commission. A license for such operation shall be issued only to an alien admitted to the United States for permanent residence who has filed under section 334(f) of the Immigration and Nationality Act (8 U.S.C. 1445(f) a declaration of intention to become a citizen of the United States and has successfully completed an examination pursuant to § 97.29.

§ 97.405 Application for license.

(a) Application for license shall be made on FCC Forms 610 and 610–C. Both forms may be obtained from the Commission's Washington, D.C., office or any of the Commission's field offices.

(b) The application forms shall be completed in full in English and signed by the applicant. The Commission may require the applicant to file additional information. Both applications must be filed in accordance with the instructions contained in §§ 97.11 and 97.41.

§ 97.407 Issuance, modification, or cancellation of license.

(a) The Commission may issue a license under such conditions, restrictions, and terms as it deems appropriate.

(b) At any time the Commission may, in its discretion, modify or cancel any license issued under this subpart. In this event, the licensee will be notified of the Commission's action by letter.

§ 97.409 Operating conditions.

(a) The alien applicant may not under any circumstances begin operation until he has received a license issued by the Commission.

(b) Except as stated in any condition the operational rules and procedure contained in Subparts A through E of this part shall be applicable.

(c) When the licensee under this subpart becomes a citizen of the United States it will not be necessary for him to notify the Commission of this fact until such time as the licensee desires to renew or modify his license. At the time the licensee becomes a citizen of the United States all procedural rights shall attach to his license and the Communications Act and Administrative Procedure Act shall be applicable regarding any request or application for, or modification, suspension, or cancellation of, any such license.

APPENDIX 1

EXAMINATION POINTS

Examinations for amateur radio operator licenses are conducted at the Commission's office in Washington, D.C., and at each field office of the Commission on the days designated by the Engineer in Charge of the office. Specific dates should be obtained from the Engineer in Charge of the nearest field office of the Commission.

Examinations are also given frequently, by appointment, at the Commission's offices at the following points:

Anchorage, Alaska.	San Diego, Calif.
Beaumont, Tex.	Savannah, Ga.
Mobile, Ala.	Tampa, Fla.

Examinations are also given at greater intervals at the places named below, which are visited for that purpose by Commission examiners from the field offices for such locations. For current schedules, exact time, place, and other details, inquiry should be addressed to the office conducting examinations at the chosen point.

QUARTERLY POINTS

Albany, N.Y.	Phoenix, Ariz.
Des Moines, Iowa.	Pittsburgh, Pa.
Fresno, Calif.	St. Louis, Mo.
Hartford, Conn.	Salt Lake City, Utah.
Knoxville, Tenn.	San Antonio, Tex.
Little Rock, Ark.	Sioux Falls, S. Dak.
Memphis, Tenn.	Syracuse, N.Y.
Nashville, Tenn.	Tulsa, Okla.
Oklahoma City, Okla.	Winston-Salem, N.C.
Omaha, Nebr.	

SEMIANNUAL

Albany, Ga.	Las Vegas, Nev.
Albuquerque, N. Mex.	Lubbock, Tex.
Bangor, Maine.	Montgomery, Ala.
Birmingham, Ala.	Portland, Maine.
Boise, Idaho.	Reno, Nev.
Columbia, S.C.	Salem, Va.
Corpus Christi, Tex.	Tucson, Ariz.
El Paso, Tex.	Wichita, Kans.
Jackson, Miss.	Williamsport, Pa.
Jacksonville, Fla.	Wilmington, N.C.

ANNUAL

Bakersfield, Calif.	Marquette, Mich.
Jamestown, N. Dak.	Rapid City, S. Dak.
Klamath Falls, Oreg.	

Because of a joint FCC-Civil Service Commission experiment, there are 99 additional cities where FCC radio operator examinations are available. This number may fluctuate from time-to-time during the experiment. The states involved in this experimental program are:

Alaska	Michigan
Hawaii (including Guam)	Montana
Idaho	Ohio
Illinois	Washington
Indiana	West Virginia
Kentucky	Wisconsin

Specific examination schedules are available from any of the Commission's district offices or from its main office in Washington, D.C.

APPENDIX 2

Extracts From Radio Regulations Annexed to the International Telecommunication Convention (Geneva, 1959)

ARTICLE 41—AMATEUR STATIONS

SECTION 1. Radiocommunications between amateur stations of different countries [1] shall be forbidden if the administration of one of the countries concerned has notified that it objects to such radiocommunications.

SEC. 2. (1) When transmissions between amateur stations of different countries are permitted, they shall be made in plain language and shall be limited to messages of a technical nature relating to tests and to remarks of a personal character for which, by reason of their unimportance, recourse to the public telecommunications service is not justified. It is absolutely forbidden for amateur stations to be used for transmitting international communications on behalf of third parties.

(2) The preceding provisions may be modified by special arrangements between the administrations of the countries concerned.

SEC. 3. (1) Any person operating the apparatus of an amateur station shall have proved that he is able to send correctly by hand and to receive correctly by ear, texts in Morse code signals. Administrations concerned may, however, waive this requirement in the case of stations making use exclusively of frequencies above 144 MHz.

(2) Administrations shall take such measures as they judge necessary to verify the technical qualifications of any person operating the apparatus of an amateur station.

SEC. 4. The maximum power of amateur stations shall be fixed by the administrations concerned, having regard to the technical qualifications of the operators and to the conditions under which these stations are to work.

SEC. 5. (1) All the general rules of the Convention and of these Regulations shall apply to amateur stations. In particular, the emitted frequency shall be as stable and as free from spurious emissions as the state of technical development for such stations permits.

(2) During the course of their transmissions, amateur stations shall transmit their call sign at short intervals.

[1] As may appear in public notices issued by the Commission.

RESOLUTION NO. 10

Relating to the use of the bands 7000 to 7100 kHz and 7100 to 7300 kHz by the Amateur Service and the Broadcasting Service.

The Administrative Radio Conference, Geneva, 1959,

Considering—

(a) That the sharing of frequency bands by amateur, fixed, and broadcasting services is undesirable and should be avoided;

(b) That it is desirable to have worldwide exclusive allocations for these services in Band 7;

(c) That the band 7000 to 7100 kHz is allocated on a worldwide basis exclusively to the amateur service;

(d) That the band 7100 to 7300 kHz is allocated in Regions 1 and 3 to the broadcasting service and in Region 2 to the amateur service;

resolves,

that the broadcasting service should be prohibited from the band 7000 to 7100 kHz and that broadcasting stations operating on frequencies in this band should cease such operation;

and noting,

the provisions of No. 117 of the Radio Regulations;

further resolves,

that interregional amateur contacts should be only in the band 7000 to 7100 kHz and that the administrations should make every effort to ensure that the broadcasting service in the band 7100 to 7300 kHz, in Regions 1 and 3, does not cause interference to the amateur service in Region 2; such being consistent with the provisions of No. 117 of the Radio Regulations.

APPENDIX 3 CLASSIFICATION OF EMISSIONS

For convenient reference the tabulation below is extracted from the classification of typical emissions in Part 2 of the Commission's Rules and Regulations and in the Radio Regulations, Geneva, 1959, and it includes only those general classifications which appear most applicable to the Amateur Radio Service.

Type of modulation	Type of transmission	Symbol
Amplitude	With no modulation	AØ
	Telegraph without the use of modulating audio frequency (by on-off keying).	A1
	Telegraphy by the on-off keying of an amplitude modulating audio frequency or audio frequencies or by the on-off keying of the modulated emission (special case: an unkeyed emission amplitude modulated).	A2
	Telephony	A3 [1]
	Facsimile	A4
	Television	A5
Frequency (or phase).	Telegraphy by frequency shift keying without the use of a modulating audio frequency.	F1
	Telegraphy by the on-off keying of a frequency modulating audio frequency or by the on-off keying of frequency modulated emission (special case: an unkeyed emission frequency modulated).	F2
	Telephony	F3
	Facsimile	F4
	Television	F5
Pulse		P

[1] (In Part 97) unless specified otherwise, A3 includes single and double sideband with full, reduced, or suppressed carrier.

Appendix 4

Convention Between the United States of America and Canada, Relating to the Operation by Citizens of Either Country of Certain Radio Equipment or Stations in the Other Country
(Effective May 15, 1952)

ARTICLE III

It is agreed that persons holding appropriate amateur licenses issued by either country may operate their amateur stations in the territory of the other country under the following conditions:

(a) Each visiting amateur may be required to register and receive a permit before operating any amateur station licensed by his government.

(b) The visiting amateur will identify his station by:

(1) *Radiotelegraph operation.* The amateur call sign issued to him by the licensing country followed by a slant (/) sign and the amateur call sign prefix and call area number of the country he is visiting.

(2) *Radiotelephone operation.* The amateur call sign in English issued to him by the licensing country followed by the words, "fixed," "portable" or "mobile," as appropriate, and the amateur call sign prefix and call area number of the country he is visiting.

(c) Each amateur station shall indicate at least once during each contact with another station its geographical location as nearly as possible by city and state or city and province.

(d) In other respects the amateur station shall be operated in accordance with the laws and regulations of the country in which the station is temporarily located.

Appendix 5

DETERMINATION OF ANTENNA HEIGHT ABOVE AVERAGE TERRAIN

The effective height of the transmitting antenna shall be the height of the antenna's center of radiation above "average terrain." For this purpose "effective height" shall be established as follows:

(a) On a U.S. Geological Survey Map having a scale of 1:250,000, lay out eight evenly spaced radials, extending from the transmitter site to a distance of 10 miles and beginning at (0°, 45°, 90°, 135°, 180°, 225°, 270°, 315° T.) If preferred, maps of greater scale may be used.

(b) By reference to the map contour lines, establish the ground elevation above mean sea level (AMSL) at 2, 4, 6, 8, and 10 miles from the antenna structure along each radial. If no elevation figure or contour line exists for any particular point, the nearest contour line elevation shall be employed.

(c) Calculate the arithmetic average of these 40 points of elevation (5 points of each of 8 radials).

(d) The height above average terrain of the antenna is thus the height AMSL of the antenna's center of radiation, minus the height of average terrain as calculated above.

NOTE 1: Where the transmitter is located near a large body of water, certain points of established elevation may fall over water. Where it is expected that service would be provided to land areas beyond the body of water, the points at water level in that direction should be included in the calculation of average elevation. Where it is expected that service would not be provided to land areas beyond the body of water, the points at water level should not be included in the average.

NOTE 2: In instances in which this procedure might provide unreasonable figures due to the unusual nature of the local terrain, applicant may provide additional data at his own discretion, and such data may be considered if deemed significant.

Appendix III

PROPOSED RESTRUCTURING OF THE AMATEUR RADIO SERVICE

FCC 74-1336
25225

Before the
FEDERAL COMMUNICATIONS COMMISSION
Washington, D. C. 20554

In the Matter of

Docket No. 20282

Amendment of Part 97 of the Commission's Rules concerning operator classes, privileges and requirements in the Amateur Radio Service.

RM-1016, RM-1363, RM-1454
RM-1456, RM-1516, RM-1521
RM-1526, RM-1535, RM-1568
RM-1572, RM-1602, RM-1615
RM-1629, RM-1633, RM-1656
RM-1724, RM-1793, RM-1805
RM-1841, RM-1920, RM-1947
RM-1976, RM-1991, RM-2030
RM-2043, RM-2053, RM-2149
RM-2150, RM-2162, RM-2166
RM-2216, RM-2219, RM-2256
RM-2284, RM-2449

NOTICE OF PROPOSED RULE MAKING

Adopted: December 4, 1974; Released: December 16, 1974

By the Commission: Commissioner Quello absent.

1. The Commission has before it the above listed petitions (also listed in more detail in Appendix I) for rulemaking. Principally, petitioners are seeking amendment to the Rules for the Amateur Radio Service regarding operator classes, requirements, and privileges. Some desire additional privileges for only one specific operator license class, or desire lower requirements for one specific class. Others want more extensive amendments, such as the deletion, or addition, of an entire license class. Some would establish a new "Hobby" operator license class, having no telegraphy skill requirement. Of these, RM-1841, RM-1991, and RM-2053 would have this operator class in the Citizens Radio Service. Since operation of a radio station as a hobby or diversion, i.e., an activity in and of itself[1], is prohibited in the Citizens Radio Service, we consider such operation to be one more suitable to the Amateur Radio Service. Thus, these three petitions are included in this proceeding. RM-1633, RM-1656, RM-1793, and RM-1841 are also included in Docket 19759, but will be considered herein to the extent applicable. Additionally, petitions RM-1947 and RM-2256 contain proposals otherwise pertaining to operator privileges and are included herein for that reason.

[1]See Sec. 95.83 (a)(1)

2. RM-1629 relates to the possibility for conducting operator examinations at places other than regular Commission examination points by persons other than Commission employees. Since the entire matter of amateur radio operator examinations will be under consideration in this proceeding, it is also incorporated.

3. The type of amendments requested by the petitioners cover a broad scope of thoughts and ideas. In summary, the salient requests are:

a. Authorize some, or all, Novice Class privileges to the Technician Class.
b. Permit a person to hold both a Novice Class license and a Technician Class license.
c. Authorize some privileges in the 144-148 MHz frequency band to the Novice Class.
d. Authorize all of the 144-148 MHz frequency band to the Technician Class.
e. Authorize some privileges in the 28-29.7 MHz frequency band to the Technician Class.
f. Reallocate the frequency subbands among the various license classes.
g. Establish new frequency subbands for incentive purposes in the 1800-2000 kHz band.

h. Authorize Amateur Extra Class operator privileges to Advanced Class operators.
i. Limit transmitter power privileges for General Class operators to 250 watts on the 3.5 MHz, 7.0 MHz and 14.0 MHz frequency bands.
j. Limit transmitter power privileges for all operator classes to 300 watts on amateur frequency bands below 30 MHz.
k. Specify maximum transmitter power in terms of output.
l. Establish a new Hobby Class license, or a new VHF Telephony Class license having no telegraphy requirements or privileges.
m. Establish a new Beginner Class and a new Code Class of operator licenses.
n. Combine the Novice Class license and the Technician Class license into a new VHF Telephony Class.
o. Establish a new Intermediate Class license and a new Communicator Class license.
p. Establish a new Advanced Technician or First Class Technician Class license.
q. Discontinue the Conditional Class and Technician Class operator license.
r. Issue the Amateur Extra Class operator license for life.
s. Reduce Element 1 (B) telegraphy requirement from 13 words per minute to 10 words per minute.

Obviously, we cannot accommodate all of these requests because some are in conflict with others. We do not believe it is desirable to deal with these petitions on a piecemeal basis, since many are interrelated. Accordingly, we conclude the time is propitious for a review of our entire amateur licensing structure. To this end, we have reviewed the petitions carefully, together with the existing system of operator privileges and requirements, against the fundamental basis and purpose of the Amateur Radio Service. The following represents our best forecast of the direction we should move in this matter.

4. We recognize the desire by some amateurs, and would be amateurs, as expressed in RM-1633, RM-1793, and RM-1976, for a class of amateur operator license having requirements that do not include a knowledge of telegraphy. Although every amateur radio operator license has traditionally required the applicant to demonstrate some level of proficiency in International Morse Code, goals within the basis and purpose of the Service could be met, at least in part, without this requirement. Moreover, as several of the petitioners point out, the International Radio Regulations do allow the Commission to waive the requirement for an amateur to " . . . . have proved that he is able to send correctly by hand and to receive correctly by ear, texts in Morse code signals . . . . ."[2] in the case of stations only operated above 144 MHz. A survey and analysis[3] conducted in 1971 indicated that there may be as many non-licensees interested in amateur radio activities, if not more, than there are persons already licensed in the Amateur Radio Service. The most often mentioned reason for not obtaining an amateur license is the telegraphy requirements. We are aware the need for, and the use of, telegraphy in amateur radio communications is much less on amateur frequency bands above 50 MHz than it is on the amateur frequency bands in the High Frequency (3-30 MHz) and Medium Frequency (.3-3 MHz) range, where spectrum conservation, tolerance to interference, and other factors, make telegraphy an important mode of amateur radio-communication. We believe, under carefully established provisions, a new "telephony-only" type of operator license, limited to frequencies above 144 MHz, could and should be incorporated into the Amateur Radio Service.

[2]Radio Regulation Annexed to the International Telecommunication Convention (Geneva, 1959) Article 41, Section 3(1).

[3]A Survey and Analysis of the Citizens Radio Service. P.B. - 204 595

5. The present operator license structure is shown in Figure 1. For all intents and purposes there are ten classes of operator licenses available in five ascending levels of operator privileges. Qualification for an operator license is established by means of the various examination elements shown in Figure 1. These may be administered either by a Commission examiner or by a volunteer examiner through the mail examination system. The Amateur Extra (C) Class, the Advanced (C) Class, and the Conditional (P) Class licenses are issued to physically disabled applicants qualifying on the basis of a mail examination administered by a

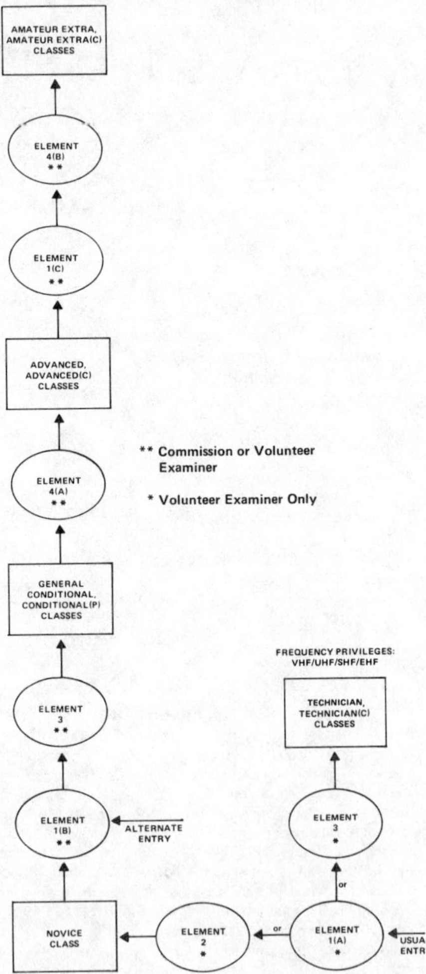

FIGURE 1. PRESENT STRUCTURE OF OPERATOR LICENSE CLASSES, EXAMINATION ELEMENTS AND FREQUENCY PRIVILEGES.

volunteer. The Conditional Class license is issued to applicants qualifying on the basis of a mail examination administered by a volunteer, because of distance or other unusual difficulty in appearing at a regular examination point. The Technician (C) Class and Novice Class licenses are issued to applicants qualifying on the basis of a mail examination administered by a volunteer, the normal procedure for these license classes. Except for the Novice Class license and the Conditional Class license, the absence of the designator (C) or (P) following the operator class on the license means the licensee has qualified before a Commission examiner, and is not subject to re-examination. Any licensee qualifying on the basis of mail examination may be required by the Commission to appear before a Commission designated examiner for re-examination. Periodically, a simple number of licensees who have obtained their licenses on the basis of a mail examination are selected at random and asked to appear in order to verify the validity of the mail examination system. Those who do not appear, and those who do not pass the re-examination, are subject to license cancellation.

6. The privileges associated with each operator license class are intended to provide the necessary incentives for amateurs to upgrade their skills. This system has been largely responsible for thousands of amateurs to upgrade, particularly to Advanced Class and to Amateur Extra Class. The current number of operators in each license class is shown in Figure 2. While it is gratifying to see even the limited success of this system toward fulfilling the basis and

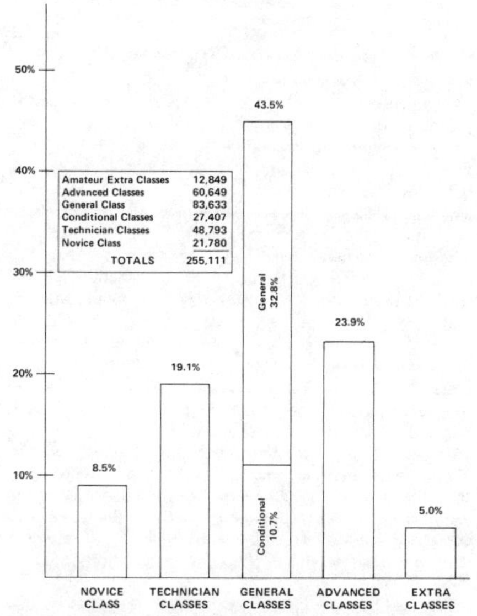

FIGURE 2. PERCENTAGE OF AMATEUR LICENSEES IN EACH CLASS, May, 1974

purpose of the Amateur Radio Service, it is a desirable goal for most amateurs to reach a higher operator class, say the Advanced Class, or even the Amateur Extra Class.[4]

7. An inherent principle in arriving at any new licensing system is a logical relationship between the qualification requirements and the operator privileges authorized at each license class level. For instance, it would not be rational to require an applicant to demonstrate a certain specific proficiency in order to qualify for a particular operator license class which authorized no corresponding privileges requiring that proficiency. Conversely, an operator license class should not authorize significantly more privileges than the requirements for that license class reasonably justify. While we believe there are the means available within the Amateur Service to satisfy the reasonable needs of most United States citizens having a genuine interest in pursuing radio activities within the basis and purpose of the Service, there are basic limitations brought about by practical realities. For example, the vast array of interests and levels of ability among amateurs must be provided for within a fixed number of different operator license classes. The resources available to the Commission for regulating the Service are not unlimited. Issuing licenses, preparing and conducting examinations, monitoring the frequencies, enforcing the regulations, etc., are all activities that must be provided by the Commission. In this proceeding, we are moving on the assumption the amateurs' record of cooperation and assistance will continue in the future, and an unduly large increase in the Commission's workload will not be necessary.

8. We are proposing in this proceeding to establish a new **Communicator Class** operator license, having no telegraphy requirements nor privileges. Operation under this license would be limited in a manner similar to that of the current Novice Class, except frequency privileges would be above 144 MHz. The objective would be to enable beginners to enter the Amateur Radio Service and, through the experience gained by operation of a low-power radiotelephony station, develop the necessary interest and skill to qualify for higher class operator licenses.

9. Those petitions calling for another new operator license class above the current Technician Class raise significant questions regarding the scope of the Technician Class as presently constituted. For example, in RM-1535, the American Radio Relay League (ARRL) states:

> "It is readily apparent from the various pronouncements of the Commission over the years and from the present interests and operations of Technician Class licenses that the purposes for which the Technician Class was established..... require review. It is respectfully suggested that any Notice of Proposed Rule Making invite comments and suggestions for major revisions of the Technician Class license...... In numerous disasters...... including the Alaskan earthquake in 1964 and the recent Hurricane Camille, the contributions of Technicians in providing internal communications have been valuable beyond estimation. Participation by Technicians in the League's Amateur Radio Emergency Corps (AREC) has grown over the years. The evolution of Technicians as communicators as well as experimenters.... since the class was established must be recognized."

Although interest in the communication aspects of amateur radio has emerged among the some 49,000 Technician Class licensees, apparently they are not sufficiently persuaded by the additional communication privileges in the High Frequency (HF) and Medium Frequency (MF) amateur bands afforded to General Class licensees to the extent of increasing their telegraphy skill from 5 words to 13 words per minute, the only real difference in qualification between the two license classes. The needs and interests of this group probably are fully satisfied by the operation of an amateur radio station in the VHF (Very-high Frequency) regions and above. Accordingly, we can conclude technological and operational developments by amateurs in the VHF, and possibly in the UHF (Ultra-high Frequency) bands, have reached the point where the interest to amateurs is comparable to, if not already exceeding, that in the MF and HF "shortwave" bands. Therefore, in order to provide meaningful incentives for amateurs interested in this part of the radio spectrum to upgrade their skills, the incentive principles should also be applied for these bands similar to those now in effect in the shortwave bands. A new higher class operator license comparable in requirements and privileges to the Advanced Class, except based

[4]Sec. 97.1(c) states as one of the principles expressing a fundamental purpose of the Amateur Radio Service: "Encouragement and improvement of the Amateur Radio Service through the rules which provide for advancing skills in both the communication and technical phases of the art."

upon operation above 29 MHz, may be desirable. Obviously, for this new higher class license, any additional telegraphy skill is not meaningful since telegraphy is not a major communication mode in these frequency bands. However, other modes, such as television, remote control, facsimile, repeaters etc., are very meaningful, and need to be emphasized. Therefore, we are proposing another new operator class license, the **Experimenter Class**, as the means toward fulfilling these needs.

10. We have examined several possible revised operator license class structures in a search for the best way to incorporate the proposed Communicator Class and the proposed Experimenter Class licenses. As broad objectives, we desire to 1) preclude, or at least minimize, any adverse impact upon presently licensed amateurs, 2) closely relate requirements to privileges for each license class, 3) provide realistic upgrading steps and incentives, 4) provide the opportunity and flexibility for persons interested only in shortwave radio, or only in VHF and above, or interested in both, to obtain a license and pursue their particular interests. As a result, the structure we are proposing is shown in Figure 3, and the specific proposed rule amendments are given in Appendix II. In general, we favor this structure because it seems to more fully reflect our objectives and to satisfy most of the objectives of the petitioners. Two series of operator license would be offered, Series A and Series B. Amateurs would be permitted to hold **one** operator license permitting privileges in one or both series. For example, an amateur could hold an operator license authorizing Novice Class privileges in Series A and also Technician Class privileges in Series B, a request asked for by several petitioners. Operator licenses in Series A would authorize only privileges on amateur frequencies below 29 MHz, and operator licenses in Series B would authorize only privileges on amateur frequencies above 29 MHz. Operator licenses would normally be issued for a 5 year renewable term, including the Communicator Class and the Novice Class in order to compensate for any increased administrative burdens resulting from the proposed amendments. (Novice Class licenses are currently issued on a 2 year, non-renewable basis, no filing fee). Section 303(L)(1) of the Communications Act of 1934 does allow us to issue operator licenses for life, as requested for the Amateur Extra Class in RM-2030. Under our current rules, the operator license[5] is always combined with the Primary station license which cannot be granted for a term longer than 5 years, a requirement of Section 307(d) of the Act. We are proposing to adopt the request. Our records indicate very few amateurs drop out of amateur radio after they have attained the Amateur Extra Class. The licensee would still be required to renew his station license(s) every five years, so in effect, this proposed rule would amount to eliminating the need to retake the examinations should the amateur neglect to renew his license.

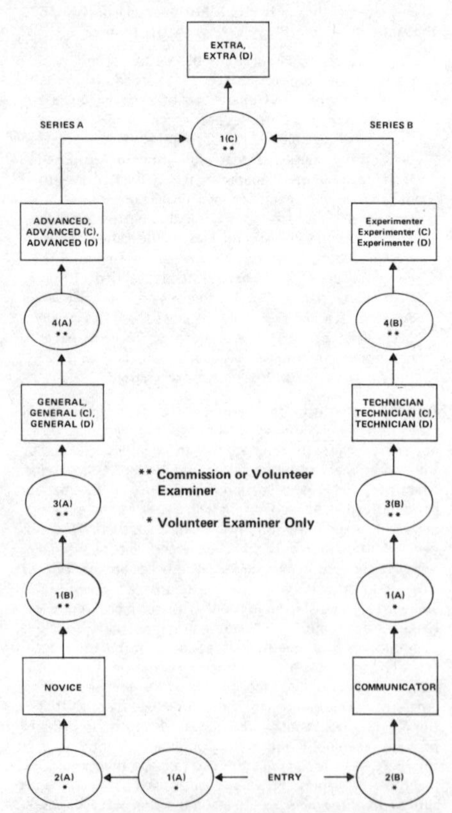

FIGURE 3. PROPOSED REVISED STRUCTURE OF OPERATOR LICENSE CLASSES AND EXAMINATION ELEMENTS

11. Under the proposed license class structure, new Advanced Class licenses and General Class licenses would no longer carry requirements and privileges above 29 MHz. The Experimenter Class and the Technician Class would be the counterpart operator licenses in Series B, and would not carry any requirements and privileges in Series A frequency bands. The current Amateur Extra Class would be shortened in name to Extra Class, and would authorize full amateur privileges in both series. We are proposing to discontinue the written examination and the exclusive telephony segments available only to this class. The material in the current Element 4(B) examination required for Amateur Extra Class would be combined with the material for the current Advanced Class Element 4(A) and, together with other new material, be used in new examination Elements 4(A) and 4(B) for the Advanced Class and Experimenter Class respectively. Material related to the shortwave domain would be used in 4(A) and material related to the other domains would be used in 4 (B). After obtaining both the Advanced Class and Experimenter Class, an amateur would then only need pass the Element 1(C) 20 word per minute telegraphy examination to qualify for the Extra Class. Because of this additional telegraphy requirement, the Extra Class would continue to have exclusive telegraphy subband privileges.

12. Under the proposed license structure, every currently licensed amateur radio operator would automatically be eligible to renew upon application, his current operator license to include privileges in at least one, and in most cases both series without further examination. Table 1 illustrates the highest class, or classes, of operator license that could be obtained without further examination.

13. Both of the proposed license series would be based upon three levels of difficulty: a beginner level, an intermediate level, and an advanced level. Ideally, this type of system would offer a newcomer the opportunity to enter the Amateur Radio Service at the beginner level with a minimum of proficiency, gain the experience and practical knowledge necessary to qualify for the intermediate level, and then move on to the advanced level. The privileges authorized at both the beginner and intermediate levels would be only those necessary to provide the desired experience for upgrading. Similarly, the related qualifica-

[5]Although large certificates are awarded to Amateur Extra Class licensees upon request, the certificates do not satisfy the availability requirements of Sec. 97.83.

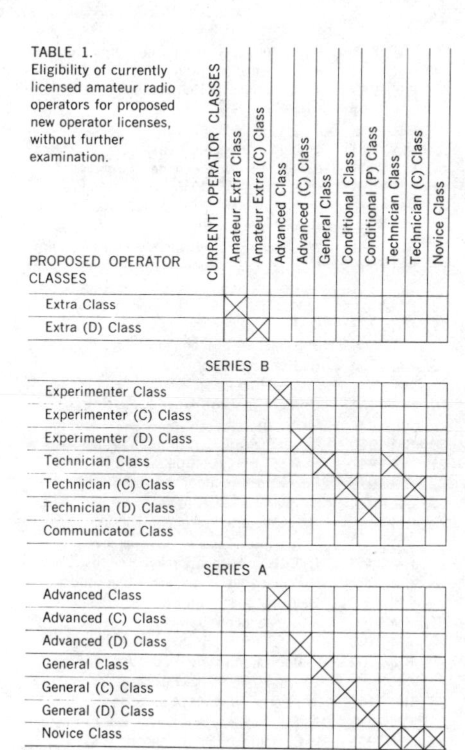

TABLE 1.
Eligibility of currently licensed amateur radio operators for proposed new operator licenses, without further examination.

PROPOSED OPERATOR CLASSES / CURRENT OPERATOR CLASSES	Amateur Extra Class	Amateur Extra (C) Class	Advanced Class	Advanced (C) Class	General Class	Conditional Class	Conditional (P) Class	Technician Class	Technician (C) Class	Novice Class
Extra Class	X									
Extra (D) Class		X								
SERIES B										
Experimenter Class			X							
Experimenter (C) Class										
Experimenter (D) Class				X						
Technician Class					X			X		
Technician (C) Class						X			X	
Technician (D) Class							X			
Communicator Class										
SERIES A										
Advanced Class			X							
Advanced (C) Class										
Advanced (D) Class				X						
General Class					X					
General (C) Class						X				
General (D) Class							X			
Novice Class								X	X	X

tion requirements would be only the minimum necessary to insure that the licensee understands the privileges, limitations, and responsibilities associated with the license, with particular emphasis on methods for properly evaluating emissions of the type(s) authorized by the license.

14. We are proposing three principal areas of operator privileges: operating frequencies, emissions, and maximum transmitter power. For Series A, the authorized frequency bands would be basically the same as at the present below 29 MHz, except the exclusive telephony segments reserved to the Amateur Extra Class would be also available to the Advanced Class. In Series B, the Technician Class would be authorized all amateur frequencies above 50 MHz, thus gaining additional frequencies 50.0-50.1 MHz and 144-145 MHz. The Experimenter Class would be authorized all above 29 MHz, and the Communicator Class all above 144 MHz. The Extra Class and Experimenter Class would be authorized all amateur emissions. The Advanced Class would be authorized all amateur emissions permitted below 29 MHz. The General and Technician Classes would be authorized emissions A1, A3, and F3. The Novice Class would continue with A1 only, while the new Communicator Class would be permitted emission F3. Related examination elements would contain questions concerning the technical and operational aspects of the emissions authorized.

15. In proposing maximum transmitter power levels, we have taken into consideration a number of factors. Amateur transmitters have not been a significant source of interference to other services, and where there has been a problem, amateurs have been very cooperative. Also, amateurs, by and large, do use the minimum transmitter power necessary to conduct their communications. Therefore, there should be no real problem if the limits were to be increased in some instances. We would like to improve the technique specified in the Rules for determining power. Modern communications requires better methods for determining transmitter power than the "plate voltage times current" method. We are proposing to specify the maximum transmitter output in terms of peak envelope power (PEP), except at the beginner level where the emissions authorized do permit a fairly accurate measurement to be made of the input power using the method now specified. Under current rules and practices, the maximum output peak envelope power that could be developed would be on the order of 2000 watts (100% modulated, full carrier, double sideband A3). Specifying this level as advanced amateur practice, and 6 dB (approximately one "S" unit) as intermediate amateur practice (500 watts PEP output) is the method used to establish these proposals. An additional 6 dB lower step (250 watts input approximates 125 watts output PEP for A1 and F3 emissions) would be the beginner level.[6]

16. The requirements for a new license, as shown in Appendix II, are similar to those now in effect, except the content of the various examination elements would be adjusted to more closely correlate with the privileges for each particular license class. While we are not proposing to lower, or increase, the telegraphy speed requirements, we are proposing a modification in the manner of testing. In RM-1724, the petitioner claims most operators must pass through a "code hump" between the speeds of 11 and 13 words per minute. Possibly the 5 wpm rate and the 10 wpm rate require the same skill level. In any event, the 13 wpm rate does require a skill level above that required for a 5 wpm rate. Otherwise, there would be no point to have both a Technician Class and a General Class under the present rules, since the two skill levels are the only difference between the requirements. Therefore, we are not proposing any changes in the telegraphy examination speeds.

17. Under the proposed system, the operator license for an amateur qualifying by means of a mail examination on the basis of a protracted physical disability would have the letter (D) inserted following the operator class [example: Advanced (D) class]. A license of this type would be renewable without re-examination upon satisfactory showing the disability continued, and they could not appear for a regular Commission supervised examination. Otherwise, they would be required to demonstrate their proficiency through re-examination. The operator license of an amateur qualifying by means of a mail examination on the basis of difficulty in traveling to a regular Commission examination point, would have the letter (C) inserted following the operator class (example: General (C) class). The only purpose of this conditionally issued type of license would be to provide a temporary authorization until the person could qualify before a Commission examiner. Hence, these licenses would not be renewable, since it would not be unreasonable to expect a conditionally licensed amateur to travel to one of the many Commission examining points sometime within the five year period. He would then have to successfully complete a regular Commission supervised examination in order to continue as an amateur radio operator.

18. In the best interests of the Amateur Radio Service, and to be fair to all amateurs, we believe that every applicant should clearly establish his qualifications for the privileges authorized by an amateur radio operator license. Overall, our experience indicates mail examinations are not as effective as Commission supervised examinations in establishing qualifications. Because of our experience in re-examining amateur radio operators, and considering the proposed amendments may place additional demands upon a mail examination system, we are proposing some amend-

[6]See[1], page 15 of appendix II.

ments in Appendix 2 intended to improve the system. Only an Extra Class licensee would be eligible to serve as a volunteer examiner for all examination elements. Advanced Class licensees would be eligible to administer examination elements for the General (C) and (D), and Novice Classes. Experimenter Class licensees would be eligible to administer examination elements for the Technician (C) and (D), and the Communicator Classes. Another proposal is to increase the required number of persons administering a volunteer examination. The second person may be the holder of any class of amateur operator license.

19. A specific call sign proposal is not included in this proceeding. However, because of the ramifications of this proposal, some relative comments are appropriate. **Existing licensees** will be able to retain current call signs if desired, and if authorized for both privileges, the same call sign may be used in both Series A and B. Licensees in Series B entering amateur radio as a result of this proceeding, will be issued a distinctive call sign for operation in that Series. If a later authorization for Series A privileges is granted, the single resulting call sign will reflect the dual Series authorization. Under this proposal, Technician Class licensees could obtain Novice privileges in Series A without examination, and therefore could retain their present call signs if desired. Further details will be contained in the call sign proceeding to be issued.

20. In view of the extensive amendments to the rules requested by the petitioners, and those proposed herein, it is imperative those submitting comments carefully consider the future needs of the Amateur Radio Service. To this end, we are allowing more than the normal amount of time for suggestions and comments to be filed. These proposals represent our best thoughts in these important matters. We are interested in receiving comments from informed amateurs in these areas.

21. Authority for the proposed rule changes herein is contained in Sec. 4(i) and 303 of the Communications Act of 1934, as amended.

22. Pursuant to applicable procedures set forth in Sec. 1.415 of the Commission's Rules, interested persons may file comments on or before June 16, 1975, and reply comments on or before July 16, 1975. All relevant and timely comments and reply comments will be considered by the Commission before final action is taken in this proceeding. In reaching its decision on the rules which are proposed herein, the Commission may also take into account other relevant information before it, in addition to the specific comments invited by this Notice.

23. In accordance with the provision of Sec. 1.419 of the Commission's Rules and Regulations, an original and 14 copies of all comments, pleadings, briefs, or other documents shall be furnished the Commission.

24. All filings in this proceeding will be available for examination by interested parties during regular business hours in the Commission's public reference room at its headquarters in Washington, D.C., (1919 M Street, N.W.).

Attachments

Vincent J. Mullins Secretary

APPENDIX I Petitioners

1. RM-1016 D. McGarrett, Centerreach, New York
2. RM-1363 K. J. Deskur, Endwell, New York
3. RM-1454 S. C. Davis, Manchester, Connecticut
4. RM-1456 W. Green, Peterborough, New Hampshire
5. RM-1516 E. W. DeCloedt, Cupertino, California
6. RM-1521 W. A. Welch, II, Wappong, Connecticut
7. RM-1526 E. C. Lips, Pittsburgh, Pennsylvania
8. RM-1535 American Radio Relay League, Newington, Connecticut
9. RM-1568 E. E. Gooch, Brilliant, Ohio
10. RM-1572 C. DeWitt, Omaha, Nebraska
11. RM-1602 C. R. Clark, Notre Dame, Indiana
12. RM-1615 C. C. Drumeller, Warr Acres, Oklahoma
13. RM-1629 M. K. Gormley, APO, New York, New York
14. RM-1633 W. Green, Peterborough, New Hampshire
15. RM-1656 Ronald A. Reed, West Los Angeles, California
16. RM-1724 R. A. Cowan, Port Washington, New York
17. RM-1793 G. Jacobs, Silver Springs, Maryland, S. F. Meyer, Linden, New Jersey
18. RM-1805 Radiotrician Confederation, Grouse Creek, Utah
19. RM-1841 United CB'ers of America, Detroit, Michigan
20. RM-1920 C. W. Tazewll, Baltimore, Maryland
21. RM-1947 R. R. Dopmeyer, Opelousa, Louisiana
22. RM-1976 Edgewood Amateur Radio Society, Baldwin Park, California
23. RM-1991 U. S. Citizens Radio Council
24. RM-2030 L. E. White, Closter, New Jersey
25. RM-2043 R. E. Heimberger, Shaker Heights, Ohio
26. RM-2053 Hercules Radio and Recording Studio, Daytona Beach, Florida
27. RM-2149 M. R. Wardean, Venice, California
28. RM-2150 W. A. Schroeder, Cherry Hill, New Jersey
29. RM-2162 Falmouth Amateur Radio Association, Woods Hole, Massachusetts
30. RM-2166 W. Brady, Norwalk, California
31. RM-2216 H. M. Krawetz, Sunnyvale, California
32. RM-2219 J. C. Hallford, Ft. Stockton, Texas
33. RM-2256 M. S. Donnell, San Jose, California
34. RM-2284 S. E. Green, et al, Austin, Texas
35. RM-2449 P. Williams, Santa Cruz, California

APPENDIX II

Part 97, of Chapter I of Title 47 of the Code of Federal Regulations is amended as follows:

1. Sec. 97.5 is amend to read:
Sec. 97.5 **Classes of operator licenses.**

(a) The following Series A operator licenses authorize operations in the amateur radio frequency bands below 29 MHz:

(1) **Advanced Class, Advanced (C) Class, Advanced (D) Class.**
Licenses to conduct amateur radio communications using advanced level amateur practices.

(2) **General Class, General (C) Class, General (D) Class.** Intermediate grade licenses to conduct amateur radio communication for the purpose of developing individual proficiency toward qualifying for the Advanced Class license.

(3) **Novice Class.** Introductory grade license to conduct amateur radio operation for the purpose of developing proficiency toward qualifying for the General Class license.

(b) The following Series B operator licenses authorize operations in the amateur radio frequency bands above 29 MHz:

(1) **Experimenter Class, Experimenter (C) Class, Experimenter (D) Class.** Licenses to conduct amateur radio communication using advanced level practices.

(2) **Technician Class, Technician (C) Class, Technician (D) Class.** Intermediate grade licenses to conduct amateur radio communication for the purpose of developing individual proficiency toward qualifying for the Experimenter Class license.

(3) **Communicator Class.** Introductory grade license to conduct amateur radio communication for the purpose of developing individual proficiency toward qualifying for the Technician Class and Novice Class licenses.

(c) The Extra Class and Extra (D) Class licenses authorize amateur radio operation using all authorized privileges, including certain exclusive privileges.

(d) The designator (C) following the type of operator license class indicates the license is conditionally issued because the licensee qualified under the provisions of Sec. 97.28.

(e) The designator (D) following the type of operator license class indicates the license is conditionally issued because the licensee qualified under the provisions of Sec. 97.27.

2. Sec. 97.7 is amended to read as follows:

Sec. 97.7 **Privileges of operator license.**

The following operating privileges are authorized by the class of operator license indicated for all new amateur licenses issued after (effective date of new rules). Amateurs licensed prior to the date will receive a new license upon the first renewal after (effective date of new rules).

(a) **Extra Classes.** All amateur radio operator privileges.

(b) **Advanced Classes.** All amateur radio operator privileges below 29 MHz, except for frequencies 3500-3525 kHz, 7000-7025 kHz, 14000-14025 kHz, and 21.000-21.025 MHz.

(c) **General Classes.**

(1) Frequencies 1800-2000 kHz, 3525-3775 kHz, 3890-4000 kHz, 7.025-7.150 MHz, 7.225-7.300 MHz, 14.025-14.200 kHz, 14.275-14.350 kHz, 21.025-21.250 MHz, 21.350-21.450 MHz and 28.0-29.0 MHz within the limitations of Sec. 97.61.

(2) Emissions A1, A3, and F3.

(3) Except for power limitations set forth in Sec. 97.61, the maximum transmitter output power shall not exceed 500 watts peak envelope power.

(d) **Novice Class.**

(1) Frequencies 3700-3750 kHz, 7100-7150 kHz, (7050-7075 kHz when the amateur radio operation is not within Region 2), 21.100-21.200 MHz, and 28.100-28.200 MHz.

(2) Emission A1.

(3) 250 watts input power to the transmitter final amplifying stage supplying radio frequency energy to the antenna, exclusive of power for heating the cathode of a vacuum tube(s), within the limitations of Sec. 97.67.

(e) **Experimenter Classes.** All amateur radio operator privileges above 29 MHz.

(f) **Technician Classes.**

(1) All amateur frequencies above 50 MHz.

(2) Emissions A1, A3, and F3.

(3) Except for power limitations set forth in Sec. 97.61, the maximum transmitter output power shall not exceed 500 watts peak envelope power.

(g) **Communicator Class.**

(1) All amateur frequencies above 144 MHz.

(2) Emission F3.

(3) 250 watts input power to the transmitter final amplifying stage supplying radio frequency energy to the antenna, exclusive of power for heating the cathode of a vacuum tube(s), within the limitations of Sec. 97.67.

3. Sec. 97.9 is revised to read as follows:

Sec. 97.9 **Eligibility for a new operator license.**

Any citizen[1] or national of the United States is eligible to apply for an amateur radio operator license. A person may be issued no more than one operator license in Series A, and no more than one in Series B. A holder of an Extra Class operator license may not hold any other amateur radio operator license issued by the Commission. The requirements for each operator class are:

(a) **Extra Class**: Applicant shall have successfully completed examination elements 1(C), 2(A), 2(B), 3(A), 3(B), 4(A), and 4(B).

(b) **Advanced Class**: Applicant shall have successfully completed examination elements 1(B), 2(A), 3(A), and 4(A).

(c) **General Class**: Applicant shall have successfully completed examination elements 1(B), 2(A), and 3(A).

(d) **Novice Class**: Applicant shall have successfully completed examination elements 1(A) and 2(A).

(e) **Experimenter Class**: Applicant shall have successfully completed examination elements 1(A), 2(B), 3(B), and 4(B).

(f) **Technician Class**: Applicant shall have successfully completed examination elements 1(A), 2(B), and 3(B).

(g) **Communicator Class**: Applicant shall have successfully completed examination element 2(B).

4. Section 97.13 and headnote are revised to read as follows:

Sec. 97.13 **Eligibility for renewal of operator license.**

(a) An amateur radio operator license, other than a conditionally issued license, may be renewed upon proper application, in which it is stated that the applicant is fully qualified in the requirements for the original license of the class being renewed. If the applicant is not fully qualified, the license will not be renewed, and the applicant may apply for a new operator license if and when he qualifies by examination at a later date.

(b) If a license, other than a conditionally issued license, is allowed to expire, application for renewal may be made during a period of grace of 1 year after the expiration date. During this 1 year period of grace, an expired license is not valid. A license renewed during the grace period will be dated currently and will not be backdated to the date of its expiration.

(c) Application for renewal of an amateur radio operator license shall be submitted on FCC Form 610 and shall be accompanied by the applicant's operator license or photocopy thereof. Application for renewal of unexpired licenses must be made during the license term. In any case in which the licensee has, in accordance with the provisions of this section, made timely and sufficient application for renewal of an unexpired license, no license with reference to any activity of a continuing nature shall expire until such application shall have been finally determined.

(d) Operator licenses obtained on the basis of Sec. 97.28 are not renewable.

(e) Operator licenses obtained on the basis of Sec. 97.27 are not renewable unless the application is accompanied by a current physician's affidavit.

(f) Extra Class operator licenses are issued for the life of the licensee, and do not have to be renewed.

5. Section 97.15 is added new to read as follows:

Sec. 97.15 **Modification of operator license.**

(a) Application for modification of an amateur radio operator license shall be submitted on FCC Form 610 and shall be accompanied by the applicant's operator license(s) or photocopy(s) thereof.

(b) When only the name of the licensee is changed, or when only the mailing address is changed, a formal application for modification of license is not required. However, the licensee shall notify the Commission promptly of these changes. The notice, which may be in letter form, shall contain the name and address of the licensee as they appear in the Commission's records, the new name and/or address, as the case may be, the primary station call sign and class of operator license. The notice shall be sent to Federal Communications Commission, Gettysburg, Pennsylvania 17325, and a copy shall be kept by the licensee until a new license is issued.

6. Section 97.21 is revised to read as follows:

Section 97.21 **Examination elements.**

[1]Senate Bill 2457 if enacted; would delete citizenship requirement.

Examination for amateur radio operator privileges will comprise one or more of the following elements:

(a) **Element 1(A)**: Slow speed telegraphy test in International Morse code at 5 words per minute.

(b) **Element 1(B)**: Intermediate speed telegraphy test in International Morse code at 13 words per minute.

(c) **Element 1(C)**: High speed telegraphy test in International Morse code at 20 words per minute.

(d) **Element 2(A)**: Rules, basic principles, and amateur practices essential to beginners' amateur radiotelegraphy operation using the privileges authorized to the Novice Class.

(e) **Element 2(B)**: Rules, basic principles, and amateur practices essential to beginners' amateur radiotelephony operation using the privileges authorized to the Communicator Class.

(f) **Element 3(A)**: Rules, intermediate level principles, and amateur practices essential to amateur radio operation using the privileges authorized to the General Class.

(g) **Element 3(B)**: Rules, intermediate level principles, and amateur practices essential to amateur radio operation using the privileges authorized to the Technician Classes.

(h) **Element 4(A)**: Advanced level principles and amateur practices essential to amateur radio operation using the privileges authorized to the Advanced Class.

(i) **Element 4(B)**: Advanced level principles and amateur practices essential to amateur radio operation using the privileges authorized to the Experimenter Class.

7. Section 97.23 is revised to read as follows:

Sec. 97.23 **Examination requirements.**

(a) The telegraphy test required of an applicant for an amateur radio operator license shall determine the applicant's ability to send correctly by hand using a hand key (or, if supplied by the applicant, a semi-automatic or electronic, hand operated key, other than keyboard type) and to receive correctly by ear, in plain language, messages in the International Morse code at not less than the prescribed speed, counting 5 characters to the word, each numeral or each punctuation mark counting as 2 characters.

(b) All written examinations for an amateur radio operator license shall be completed by the applicant in legible handwriting or hand printing by means of ink or pencil. Whenever the applicant's signature is required, his normal signature shall be used. Applicants unable to comply with these requirements, because of a physical disability, may dictate their answers to the examination questions and to the receiving code test. If the examination, or any part thereof, is dictated by the applicant, the examiner shall certify the nature of the applicant's disability and the name and address of the person(s) taking and transcribing the dictation.

8. Section 97.25 is revised to read as follows:

Sec. 97.25 **Examination credit.**

(a) An applicant for an amateur radio operator license will be given credit for those examination elements required for any other class or operator license held when the application is filed. However, credit will not be given for examination elements 1(B), 3(A), 3(B), 4(A), and 4(B) given under the provisions of Sec. 97.30 for a class of operator license other than that being applied for, except for holders of Advanced (D) Class, Experimenter (D) Class, General (D) Class, and Technician (D) Class when qualifying for a license under the provisions of Sec. 97.27. NOTE: Credit for examination elements will be given to applicants holding a valid operator license at the time of the adoption of this rule, in accord with the following schedule, during a period not exceeding one year following the expiration date on the current license:

(1) **Amateur Extra Class**: All examination elements.

(2) **Amateur Extra(C) Class**: Elements 1(A), 2(A) and 2(B). Also all other examination elements as if passed on the basis of Sec. 97.27.

(3) **Advanced Class**: Elements 1(A), 1(B), 2(A), 2(B), 3(A), 3(B), 4(A), and 4(B).

(4) **Advanced (C) Class**: Elements 1(A), 2(A), and 2(B). Also elements 1(B), 3(A), 3(B), 4(A), and 4(B) as if passed on the basis of Sec. 97.27.

(5) **General Class**: Elements 1(A), 1(B), 2(A), 2(B), 3(A), and 3(B).

(6) **Conditional Class**: Elements 1(A), 2(A), and 2(B). Also elements 1(B), 3(A), and 3(B) if passed on the basis of Sec. 97.28.

(7) **Conditional Class (P) Class**: Elements 1(A), 2(A), and 2(B). Also elements 1(B), 3(A), and 3(B) as if passed on the basis of Sec. 97.27.

(8) **Technician Class**: Elements 1(A), 2(B), 3(A) and 3(B).

(9) **Technician (C) Class**: Elements 1(A), 2(A), and 2(B). Also elements 3(A) and 3(B) as if passed on the basis of Sec. 97.28.

(10) **Novice Class**: Elements 1(A) and 2(A).

(b) Upon request, an applicant for an amateur radio license will be given credit for element 1(A) and 1(B) if within 5 years prior to the receipt of his application by the Commission, he held a commercial radiotelegraph operator license or permit issued by the Federal Communications Commission.

(c) Upon request, an applicant for an amateur radio operator license will be given credit for elements 1(A), 1(B), and 1(C), if he holds a valid First Class commercial radiotelegraph operator license or holds any commercial radiotelegraph operator license or permit issued by the Commission containing aircraft radiotelegraph endorsement.

(d) Applicant submitting evidence of having held the Amateur Extra First Class operator license and having held its successor license will be given credit for examination element 1(C) if he so requests. An applicant must present his proof in advance of the desired examination time to the Amateur and Citizens Division, Washington, D.C., 20554 and receive a letter of certification for presentation to the Commission Field Office where the examination will be taken. No credit for the telegraphy requirement will be given without the letter of certification.

9. Section 97.27 and headnote are revised to read as follows:

Sec. 97.27 **Availability of operator license to physically disabled persons.**

If it is shown by physician's certificate an applicant is unable to travel to any regular Commission examination point because of a protracted physical disability, a new or renewed Extra (D) Class, Advanced (D) Class, Experimenter (D) Class, General (D) Class, or Technician (D) Class operator license may be issued on the basis of examinations successfully passed under the provisions of Sec. 97.30. These licenses may not be renewed without a current physician's affidavit.

10. Section 97.28 and headnote are revised to read as follows:

Sec. 97.28 **Availability of operator license to persons residing at great distances from Commission examination points.**

(a) A new Advanced (C) Class, Experimenter (C) Class, General (C) Class, or Technician (C) Class license may be issued on the basis of examinations successfully passed under the provisions of Sec. 97.30 under one of the following conditions:

(1) If the applicant's legal residence, mailing address, and/or any station location or proposed station location are more than 175 miles actual distance from the nearest Commission examining point.

(2) If the applicant is shown by certificate of the commanding officer to be in the armed forces of the United States at an Army, Navy, Air Force, or Coast Guard station and, for that reason, to be unable to appear for examination at a Commission examination point.

(3) If the applicant demonstrates by sufficient evidence that he is unable to appear at a Commission examination

point because his current temporary residence, for the 12 coming months is outside the continental limits of the United States, its territories or possessions.

(b) Operator licenses obtained under the provisions of these rules are not renewable.

11. Section 97.29 and headnote are revised to read as follows:

Sec. 97.29 **Manner of conducting Commission supervised examinations.**

(a) Except as provided by Sec. 97.27 and Sec. 97.28, examination elements 1(B), 1(C), 3(A), 3(B), 4(A) and 4(B) may only be administered by an authorized Commission employee or representative at locations and at times specified by the Commission.

(b) Examination element 4(A) may only be administered to a person having successfully passed element 3(A).

(c) Examination element 4(B) may only be administered to a person having successfully passed element 3(B).

(d) Examination element 3(A) may only be administered to a person having successfully passed element 2(A).

(e) Examination element 3(B) may only be administered to a person having successfully passed examination elements 2(B).

12. Section 97.30 is added new to read as follows:

Sec. 97.30 **Manner of conducting mail examinations**

(a) Unless otherwise prescribed by the Commission, examination elements 1(A), 2(A), 2(B), and any elements administered under the provisions of Sec. 97.27 and Sec. 97.28 will be conducted and supervised by two proxy volunteer examiners proposed by the applicant and approved by the Commission. The volunteer examiners shall be at least 21 years of age, shall be unrelated to the applicant, and at least one shall hold the proper class of license to administer examinations in accordance with the following schedule:

(1) Extra Class: All examination elements.

(2) Advanced Class: examination elements 1(A), 1(B), 2(A), and 3(A).

(3) Experimenter Class: examination elements 1(A), 2(B), and 3(B).

(b) Written examinations shall be obtained, administerd, and submitted in accordance with the following procedure:

(1) Within 10 days after successfully passing any required telegraphy examination element, an applicant shall submit an application (FCC Form 610), together with any filing fee prescribed, to the Commission's office in Gettysburg, Pennsylvania 17325. The application shall include a written request from the volunteer examiners for the appropriate examination papers. The examiners' written request shall include (1) the name and mailing address of the volunteer examiners, (2) the name of the applicant, (3) a statement by the volunteer examiners that the applicant has passed the telegraphy examination element for the class of operator license, if required, under their supervision within the 10 days prior to the submission of the request, and (4) the volunteer examiners' signatures. Examination papers will be forwarded to one of the volunteer examiners. NOTE: When the applicant is entitled to credit for any telegraphy examination element under the provisions of Sec. 97.25, an application may be submitted without regard to the 10 day limitation. The examiners' request should then state that a telegraphy examination was not administered for that reason. The applicant should furnish details as to the class, number, and expiration date of any license involved.

(2) The proxy volunteer examiners shall be responsible for the proper conduct and necessary supervisions of the examination. Administration of the examination shall be in accordance with the instructions included with the examination papers.

(3) The examination papers, either completed, or unopened in the event the examination is not administered for whatever reason, shall be returned by the volunteer examiner to the Commission's office at Gettysburg, Pennsylvania, no later than 30 days after the date the papers are mailed by the Commission (the date of mailing is normally stamped by the Commission on the outside of the examination envelope).

13. Section 97.33 is amended to read as follows:

Sec. 97.33 **Eligibility for re-examination.**

An applicant who fails an examination for an amateur radio operator license may not take another examination for the same or higher class license in the same series within 30 days.

14. Section 97.35 and headnote are revised to read as follows:

Sec. 97.35 **Additional requirements for licensees holding licenses on the basis of mail examinations.**

(a) A licensee holding an amateur radio operator license obtained by a mail examination administered by proxy volunteer examiners may be required to appear for a Commission supervised examination at a location designated by the Commission. If the licensee fails to appear for this examination when directed to do so, or fails to pass such examination, the amateur radio license(s) involved shall be subject to cancellation. When a license is cancelled under this provision, a new license will not be issued for the same class of operator license as that cancelled.

(b) A holder of an amateur radio operator license obtained on the basis of a mail examination under the provisions of Sec. 97.27 shall make application for re-examination within one-year upon becoming able to travel to any Commission examination point.

(c) A holder of an amateur radio operator license obtained on the basis of mail examination under the provisions of Sec. 97.28 shall apply for re-examination within one-year of when the licensee changes his legal residence, or mailing address, and/or any station or proposed station location within 175 miles actual distance to the nearest Commission examination point, or when a new examination point is established within 175 actual miles distance to the licensee's legal residence, mailing address, or station location.

15. Section 97.38 is added new to read as follows:

Sec. 97.38 **Types of station licenses and eligibility**

(a) The following types of station licenses are available to properly licensed amateur radio operators.

Type of station	Eligible licensees
Series A Primary station.	Extra Class, any Series A Class operator.
Series A Secondary station.	– do –
Series B Primary station.	Extra Class, any Series B Class operator.
Series B Secondary station.	– do –
Series A Club station.	Extra Class, Advanced Class operator.
Series B Club station.	Extra Class, Experimenter Class operator.
Repeater station, Control station, Auxiliary Link station, Space station.	Extra Class, Experimenter Class operator.
Military Recreation station.	Individual, whether or not a licensed amateur radio operator, who is in charge of a proposed Military Recreation station.

16. Section 97.67(a) & (b) are amended and par (d) added to read as follows:

Sec. 97.67. **Maximum authorized power.**

(a) Within all other limitations specified herein, amateur radio stations shall use the minimum amount of transmitter power necessary to carry out the desired communications.

(b) Except for power limitations set forth in Sec. 97.7 and Sec. 97.61, the maximum transmitter output power shall not exceed 2000 watts peak envelope power.[1]

(d) Any transmitter capable of exceeding the power limitations specified herein shall not be operated in the Amateur Radio Service unless there is incorporated adequate measures to insure the limitations will not be exceeded.

[1]This is one proposal under consideration. The Commission is also considering alternatives such as PEP input, average power input, ratios of peak to average power output and limitations on dissipation ratings of final power amplifier devices or a combination of these. Specific comments on the practicality of these proposals, alternate proposals and the practicality of attendant power measuring techniques by amateur stations are requested.

We request comments on the need for rules limiting the use of techniques which increase the average power in A3 single sideband supressed carrier transmissions, without increasing the peak envelope power. The comments should discuss the various techniques utilized for the purpose in the Amateur Radio Service, the engineering standards that must be observed for good amateur practice when using these techniques, the nature of any unnecessary interference that can be caused by the improper use of these techniques, and the capabilities of amateurs to make measurements necessary to proper usage.